Molecular Inclusion and Molecular Recognition Clathrates I

Clathrate Chemistry Today — Some Problems and Reflections

E. Weber

Separation of Enantiomers by Clathrate Formation

D. Worsch and F. Vögtle

Isolation and Optical Resolution of Materials Utilizin Inclusion Crystallization

F. Tode

Tri-o-Thymotide Clathrates

R. Gardil

Structural Parsimony and Structural Variety among Inclusion Compounds

F. H. Herbstein

Inclusion Properties of Tetraphenylene and Synthesis of Its Derivatives

Th. C. W. Mak and H. N. C. Wong

AKADEMIE-VERLAG BERLIN

Topics in Current Chemistry · Vol. 140

Molecular Inclusion and Molecular Recognition Clathrates I

Molecular Inclusion and Molecular Recognition Clathrates I

With 69 Figures and 40 Tables

AKADEMIE-VERLAG BERLIN
1987

Die Originalausgabe erscheint im Springer-Verlag
Berlin · Heidelberg · New York · London · Paris · Tokyo
in der Schriftenreihe "Topics in Current Chemistry",
Volume 140

Vertriebsrechte für die sozialistischen Länder:
Akademie-Verlag Berlin

Vertriebsrechte für alle Staaten mit Ausnahme der sozialistischen Länder:
Springer-Verlag Berlin · Heidelberg · New York · London · Paris · Tokyo

ISBN 3-540-17307-2 Springer-Verlag Berlin · Heidelberg · New York
ISBN 0-387-17307-2 Springer-Verlag New York · Heidelberg · Berlin

ISBN 3-05-500370-5 Akademie-Verlag Berlin

Erschienen im Akademie-Verlag Berlin, DDR-1086 Berlin, Leipziger Straße 3—4
Lizenznummer: 202 · 100/510/87
Printed in the German Democratic Republic
Gesamtherstellung: VEB Druckerei „Thomas Müntzer", 5820 Bad Langensalza
LSV 1205
Bestellnummer: 763 735 9 (3072/140)
12400

Editorial Board

Table of Contents

Preface

Chemistry is at a crossroads. In fact chemists no longer just aim at the synthesis of compounds via covalent linkage or the clarification of chemical reaction mechanisms. More and more research activities are directed to understand the nature of what is called "weak intermolecular interactions" in a broader sense. Weak non-covalent bonds involving neutral organic molecules are of vital importance, e.g. in molecular biology, drug design etc.

The search for artificial host mimics possessing specific weak interaction and complexation properties to ions and neutral molecules has inspired scientists to work with great effort during the past several years and much success can be reported. Different subareas of research related to this matter have developed (crown chemistry, cyclodextrins, calixarenes etc.). They are partly a subject of former volumes of this series (cf. Top. Curr. Chem. *98*, *101*, *121*, *123*, *125*, *128*, *132*, and *136*).

However, there is another important subject in this range of interest which has not been reviewed recently, but which is advancing so rapidly that reporting of its current state of knowledge is much desirable. The point is research on lattice-type molecular inclusion compounds or clathrates. The actual interest in these compounds is broad, in a theoretical and practical sense, ranging from the directed synthesis of new host molecules to the investigation of selective inclusion properties, molecular recognition, receptor-substrate analogy, topochemistry, X-ray crystallography, chemical analysis, molecular separation, compound protection, detoxification and other applications of potential industrial value.

This book is the first of a two-volume series intending to provide modern aspects of clathrate inclusion chemistry. Chapter 1 has the character of an introduction. It shows briefly the evolution of clathrate chemistry from the beginning in the early nineteenth century to the present and leads to the current problems. A topic which is particularly stressed in this volume is the use of clathrates in enantiomer resolution (Chapters 2–4). Some structural problems of general interest are also discussed (Chapter 5) and a new host system, tetraphenylene, is introduced in Chapter 6.

The second volume which is scheduled to appear soon will concentrate mainly on design concepts and structural aspects of new host molecules and inclusion compounds.

Bonn, November 1986 Edwin Weber

Clathrate Chemistry Today — Some Problems and Reflections

Edwin Weber

Institut für Organische Chemie und Biochemie der Universität Bonn,
Gerhard-Domagk-Str. 1, D-5300 Bonn-1, FRG

Table of Contents

1 Evolution of Crystal Inclusion (Clathrate) Chemistry — A Very Short Survey

The chemistry of inclusion compounds has a long history [1]. Confirmed accounts on the preparation of such chemical species date back to the beginning of the nineteenth century. At that early time Davy, and shortly afterwards Faraday, reported of a chlorine clathrate hydrate [2].

The chemistry of inclusion compounds also looks back on a lively history [3]. There are many events of significance in the area of inclusion chemistry till the middle of the twentieth century including the discovery of new inclusion compounds and hosts [4–8] (Fig. 1), among them the graphite intercalates, the β-quinol and cyclodextrin inclusion compounds, the Hofmann-type clathrates as well as the inclusion compounds of tri-o-thymotide, Dianin's compound, the choleic acids, of phenols, of urea and others specified in comprehensive monographs [9–12].

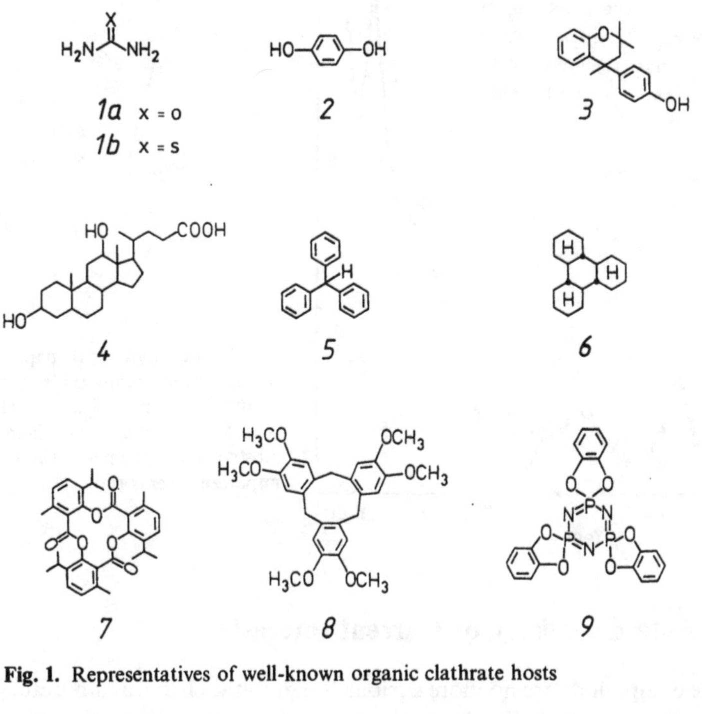

Fig. 1. Representatives of well-known organic clathrate hosts

The most important event, however, emerged from the pioneering X-ray structural work of H. M. Powell on inclusion compounds [1,13], e.g. the structural resolution of the SO_2 clathrate of hydroquinone [14]. This was about 1945 and opened-up a whole new science of the study of inclusion phenomena. Before the new tool was applied in crystallography, the nature of inclusion compounds was understood only very vaguely, but afterwards the situation changed. Arousing from this fact, the interest in all types of inclusion chemistry received an impetus of considerable extent which was only surpassed when crown hosts in 1967 [15] and a little later cryptands [16] were discovered. From an objective point of view, however, inclusion compounds formed by crowns and cryptands, e.g. with cations, do not belong to the clathrates, but are inclusion complexes [17] (see below, Sect. 4).

Irrespective if this reduction, the chemistry of "true" crystal inclusion compounds, generally spoken "clathrates", expanded enormously within the last years [18] and the tendency is still upwards [19]. Figure 2 confirms the facts by a graphical representation showing the annual distribution of the number of papers (based on references of Chemical Abstracts) on inclusion compounds between 1948 and 1981 [1]. The diagram with its overwhelming final ascent awakens one's interest in the question as follows.

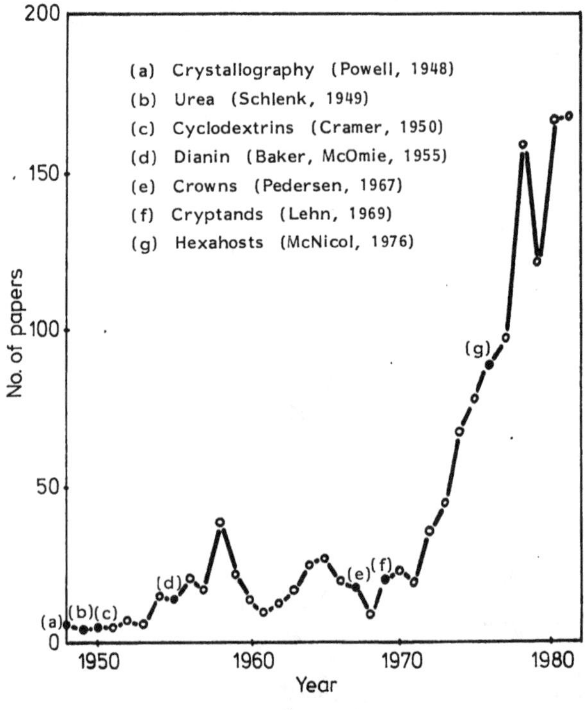

Fig. 2. Total number of papers on inclusion compounds per annum between 1948 and 1981 (based on references of Chem. Abstr.) and assignment of some important events (bold dots) [1]

2 Why is Clathrate Chemistry of Current Interest?

All along, clathrate compounds are no more curious things in the chemical laboratory as they were often understood [1]. On the contrary, there is a broad actual field of practical and research applications using clathrate compounds (Vol. 3 of Ref. 18), and a great many of future applicabilities [1,20] wait for exploration. Some few of these aspects are given in the following.

One important field of applications, industrial objectives included, is directed to *chemical analysis* and *molecular separation* processes [12]. Corresponding to the size, the shape, and the chemical nature of the holes generated in an inclusion lattice, guest molecules may be included selectively. Out of a mixture of compounds the one which matches the conditions of the lattice holes most suitably is preferrably accommodated [18]. *Chemically* different species are separated (e.g. hydrocarbons and ketones) as well as *constitutional* isomers, *positional* isomers, *regioisomers*, *stereoisomers* (*enantiomers* and *diastereomers*) and even *isotopic* isomers. In the case of enantioselective guest inclusion, a chiral host compound (chiral host lattice) is required [21].

Size selectivity which has been recognized as a characteristic feature of host lattices is an object most promising in industry [22]. Compounds which have close boiling points and therefore need costly distillation processes for separation may be separated by inclusion crystallization at a lower price. A few pilot plants involving clathrate formation have been described [12], e.g. to separate *meta*- and *para*-xylenes [23, 24]. Some few technical problems to work with inclusion compounds at a *commercial* scale do exist [25], but they can be overcome. Using the so-called "liquid clathrates" [26] which require no precipitation and filtration equipment, is another interesting alternative. On an *analytical* scale, clathrate compounds were successfully applied also in chromatography (Fig. 3) [27].

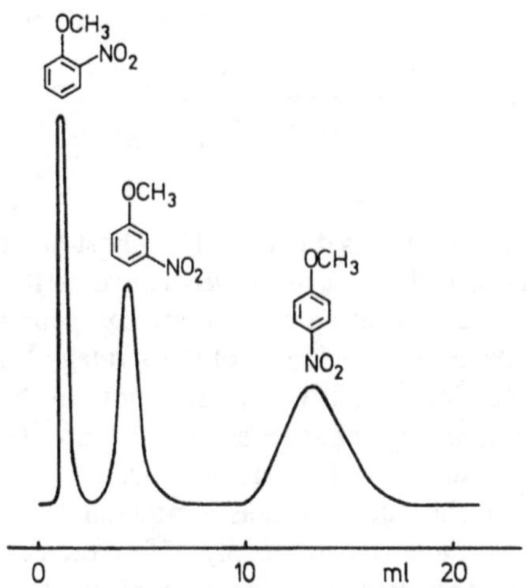

Fig. 3. Separation of an o-, m-, p-mixture of nitroanisole by a clathrate chromatography system using a Werner-type compound (elution curve) [27]

The characteristic of crystal lattices is a strict periodical succession of structurally identical molecular units, in the sense of an inclusion lattice also of holes, channels, layers etc. which may include guest molecules in an oriented fashion. This organizing principle makes *topochemistry* [28, 29] possible. One of the early studies in this area was the *inclusion polymerization* of dienes in the channels of urea, respectively thiourea, leading to stereoregular polymers (Eq. 1) [30]. Although stereodifferentiating inclusion polymerization/co-polymerization has been performed in other host lattices, too, e.g. in the channels of the perhydrotriphenylene host (*6*) [31], it is still a problem of actual interest [32].

The "chemical" reactivity of a substrate molecule, e.g. at *photo*- (Eq. 2) or *thermal isomerization* (Eq. 3) may also be altered on lattice enclosure since conformations different to the substrate free of constraints of the lattice environment are likely [33, 34]. This will lead to modified reaction path ways and thus cause different products, or if the reactivity of an included molecule is drastically reduced, the host lattice offers a *protective function*. In the simplest case the compound is protected against the influence of light and heat.

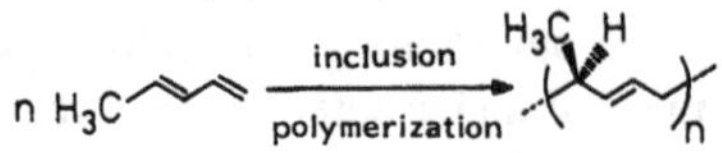

(1)

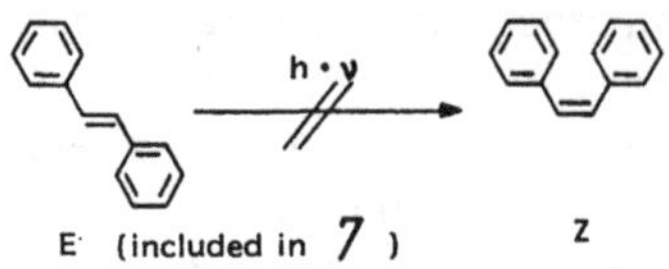

(2)

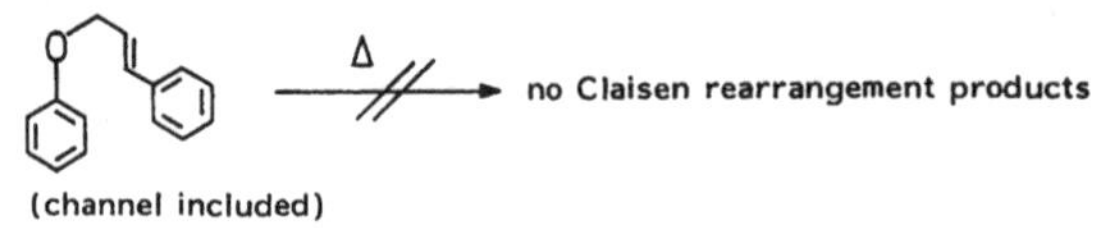

(3)

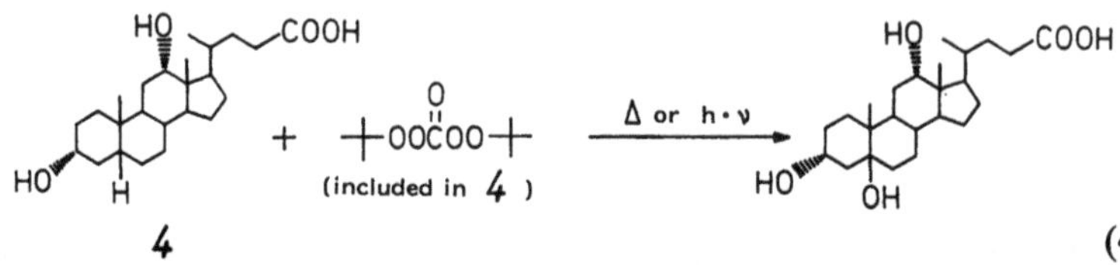

(4)

Another object which is of benefit to chemical synthesis is derived from host-guest reactivity under oriented conditions. The *regioselective* and *stereospecific* hydroxylation of deoxycholic acid (*4*) (host lattice) effected by heating its clathrate compound with di-t-butyldiperoxycarbonate (Eq. 4) is a promising beginning of this strategy [35]. Inclusion compounds may also be used as *chemical reagents* more and more in the future. Clearly the intercalates, e.g. of graphite [36] and of sheet silicates [37], are the representative examples, but there are more possibilities of development [1].

Resembling the use of the matrix isolation technique, inclusion compounds have been recognized as sources and reservoirs of *unstable species*, mainly of free radical-type [38, 39]. As mentioned before, molecules being blocked in an unusual conformation is a second facility [40–42].

In general, lattice inclusion will cause altered *physical properties* [43] of any guest molecule. This may manifest itself in a reduced volatility and therefore lower possible storage and handling problems of a compound when included [6, 44]; toxical and hazardous substances become safer [45]. Enclosure does also change the redox properties of a compound, its color and other physical dimensions [1, 10]. There is considerable interest in the altered physical properties of inclusion compounds for the use in battery systems, room temperature superconductors and further aims of future technology [46].

Studying the nature of inclusion compounds requires special *determination techniques* which will work in the solid state. Those have been developed in the range of IR- [47], Raman- [47], and magic angle spinning spectroscopy [48] as well as in high resolution electron microscopy [49], and X-ray crystallography [50] and were successfully applied on clathrates. The clathrate phenomenon, thus, acts also as a stimulus to develop physical measuring systems to a high standard. And last not least, measurements of individual spectrophysical data of the included molecules are feasible under oriented conditions.

The given applications, though only outlined by a few examples, are sufficient to illustrate what renders the clathrates and analogous types of inclusion compounds so attractive.

3 Directed Host Design versus Discovery by Chance

With regard to the many applications, general guiding lines for the easy construction of new clathrate compounds are much desirable [1, 19]. Actually most of the classical clathrate hosts [8–12] (see Fig. 1) have been discovered by accident and not via directed synthesis [20]. The only starting point of a general way of looking at clathrates and their molecular construction principles untill quite recently came from Powell's fundamental crystallographic work [13]. In this connection he recognized that the *phenolic hydroxyl* function, e.g. as present in the different quinols (cf. *2*) or the Dianin's compound (*3*), is a certain structural prerequisite being favorable to manage a host lattice [51] (Fig. 4). More of such individual building elements should exist, but which?

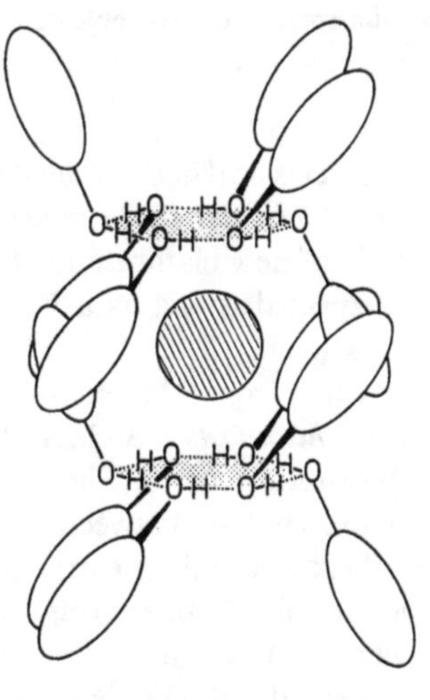

Fig. 4. Inclusion matrix of Dianin's compound (*3*) (schematic representation) [6]. Individual Dianin molecules are represented by a specified hydroxy group attached to an ellipsoid. The characteristic hydrogen bridge networks are indicated by the shaded hexagons (H-bonds in dotted lines). The hatched sphere in the centre of the cavity pictures an included guest molecule, e.g. chloroform

In the sequel, efforts were made to keep on with such ideas, i.e. creating new host compounds simply by altering an individual section of a known host constitution. Typical examples are the *Dianin's compound* (*3*) and its structural modifications [52] (Fig. 5). On the one hand it is possible to vary the cavity size being formed in the crystal lattice of Dianin's compound in a very special way via modifications of molecular segments, e.g. going from the shape of a "hour-glass" (Fig. 5a) to a "Chinese lantern" [53] (Fig. 5b). This modification involves the addition of only one simple methyl group at the 8-position in the sulphur-analogous Dianin's compound (change from *11* to *12*). One the other hand, changing the structure of Dianin's compound (*3*) to the

3′-methyl-substituted analogue (*10*) causes a complete loss of the original clathrate activity [53]. Replacement of an O in Dianin's compound by S in the sulphur analogue obviously has a drastic effect on the inclusion formation, but strange to say, only in respect to the methyl derivatives (*10* and *12*); the basic constituents (*3* and *11*), however, are much lower affected by that operation.

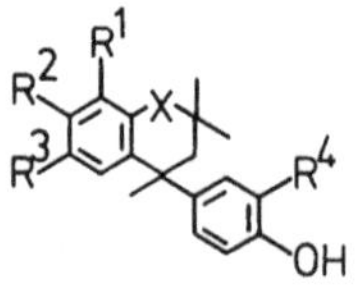

3 X = O, $R^1 - R^4 = H$
10 X = O, $R^1 - R^3 = H$, $R^4 = CH_3$
11 X = S, $R^1 - R^3 = H$ (a)
12 X = S, $R^1 = CH_3$, $R^2 - R^4 = H$ (b)

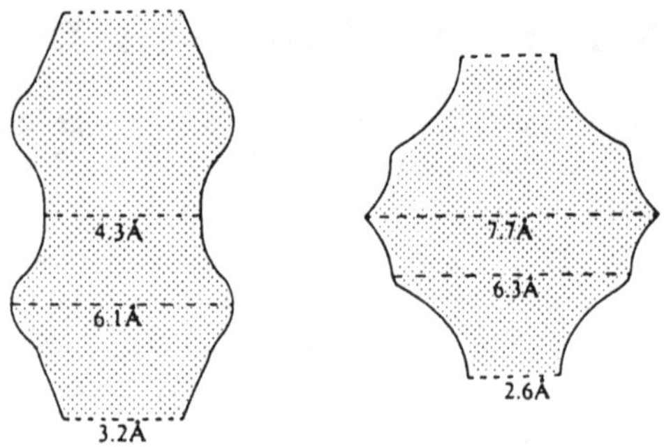

a **b**

Fig. 5. Dianin's compound (*3*) and constitutional modifications *10–12*; (**a**) and (**b**) showing the cavity contours (section through the van der Waals surface) of the lattice inter-space made available of *11* and *12*, respectively [52, 53]

It is true: the method of modifying an individual segment of a classical host molecule will not solve the real problem being connected to a *directed host design*. Directed design of a host compound, to be exact, means the synthesis of new clathrate hosts *unrelated* to any known host lattice but which would be expected to act as a host lattice. This unsatisfactory situation went on till the recent past.

The first true rational design of a new clathrate family came from MacNicol in the middle of the seventies [54] and is based on the so-called "*hexahosts*", a class of compounds featured by a hexa-substituted benzene constitution [53, 55] (Fig. 6). This original idea behind this strategy is derived from a close analogy between the hexa-substituted benzene ring and the hydrogen-bonded hexamer unit present in the clathrates of most phenol-type hosts (Dianin's compound included), bearing in mind that the O···O spacing (in the hydrogen-bonded hexagon) is similar to the distance where the atoms of a hexa-host bend off from the central unit (Fig. 6a and 6b). Choosing suitable substituents might increase the possibility to form non-close-packed structures on crystallization (Fig. 6d). Many of the compounds *13*, following this strategy, exhibit distinct inclusion ability [55].

A special merit of this host-type is that the inclusion cavity disposed is easily to tune to the geometric and steric requirements of the guest enclosure by altering the bulk of the side arms. However, guest inclusion showing a distinct specifity against different classes of compounds, i.e. chemoselectivity control, are not feasible on a desirable scale.

Another strategy which orients along the overall molecular shape of a host has been introduced by Toda already in 1968 [56] but was generalized only 16 years later

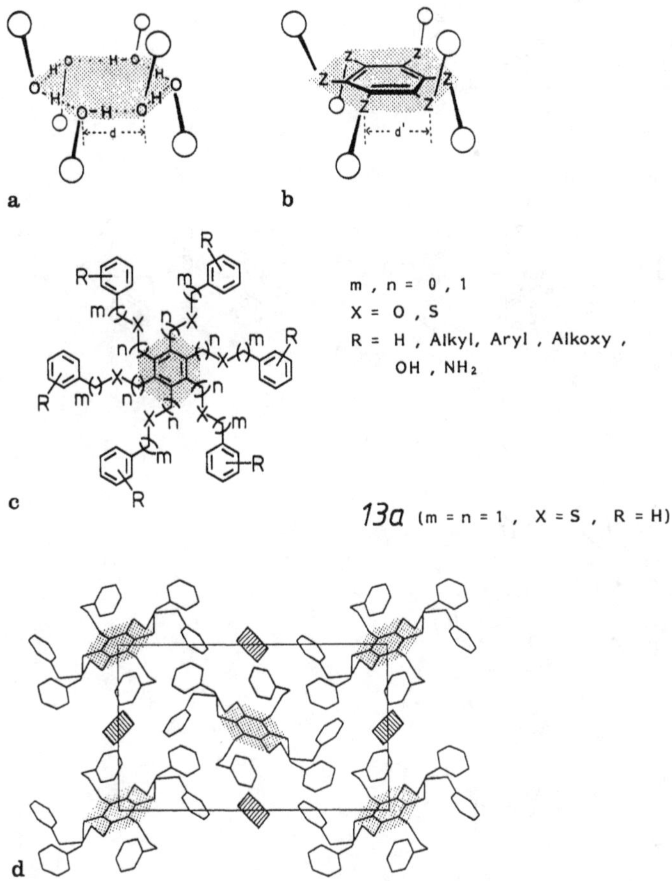

Fig. 6. "Hexa-host" strategy [53, 55]. Close connection between (**a**) a hydrogen-bonded hexamer unit typical of phenolic hosts (hydroquinone, Dianin's compound; cf. Fig. 4) and (**b**) of a hexasubstituted benzene analogue (follow up the shaded hexagons); (**c**) characteristic constitution of "hexa-host" molecules and (**d**) host-guest packing of a representative clathrate inclusion compound (dioxane clathrate of *13a*, 1:1; dioxane molecules hatched)

by the work of Hart [57]. The host design, which is spoken of, is picturally called "*wheel-and-axle*" (Fig. 7). Such hosts contain a long molecular axis made of sp carbons with sp^3 carbons at each end that bear large, relatively rigid groups (e.g. *14*, *15*). The large end groups act as "spacers" which prevent the host molecules from a close-packed structure in the crystal (cf. Fig. 7b). Hence, substantial lattice voids are created. They are well prepared to accommodate aromatic hydrocarbons according to their chemical nature, their size, and geometry. Control of the lattice dimensions is practicable by shortening, lengthening or bending of the molecular axis, e.g. using more than one sp carbons or incorporation of sp^2 and sp^3 carbons instead (*16*).

Goldberg and Hart have applied the very structural principle to bulky ureas (*17*) [58, 59] which are of advantage in that they provide a polar region at the central axis being involved in H-bonding, e.g. with the guest. The original linear acetylenic hosts (*14*)

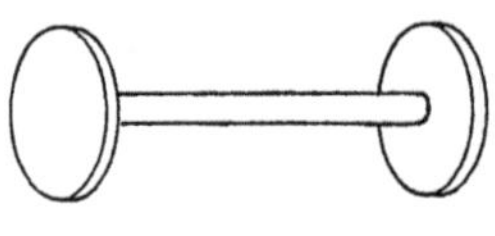

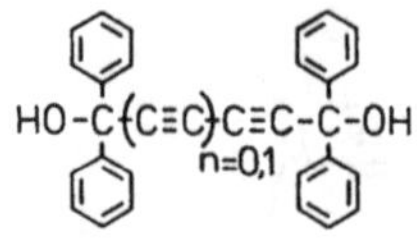

14

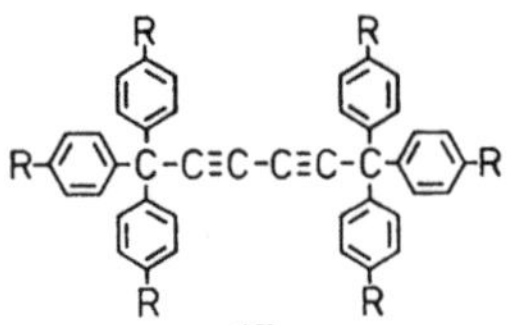

15

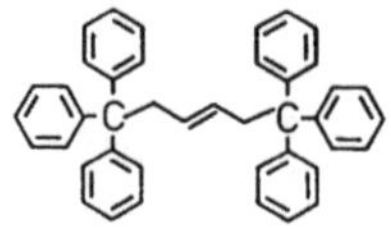

16

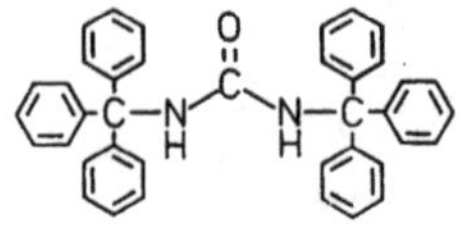

17

a R = Ph , 4-CH_3OPh

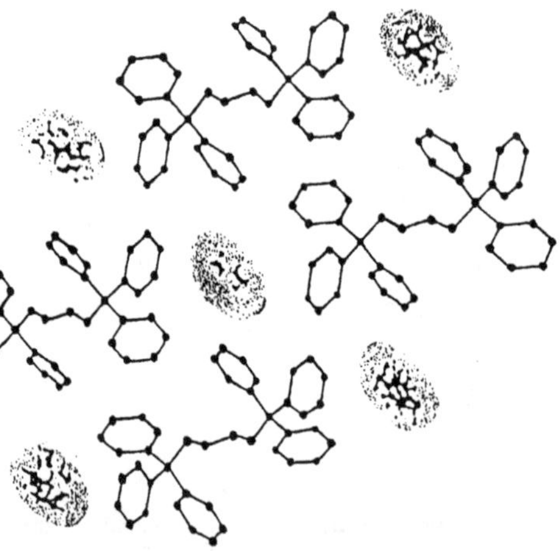

b

Fig. 7. Host design according to the "wheel-and-axle" idea: (**a**) sketch of the overall molecular shape; *14–17* showing typical examples of molecular constitutions; (**b**) packing diagram of a representative inclusion compound (*16* with toluene, 1:1; guest molecules shaded) [57)]

reported by Toda [56)] differ from Hart's development [57)] (e.g. *15*) in the presence of hydroxyl groups. Those, as a consequence of H-bonding, may also contribute to inclusion formation, e.g. in host-guest binding, if feasible. A further design of hydroxylic hosts (*18*, *19*; Fig. 8), non-related to the propargylic alcohol type of compounds, has also been applied successfully by Toda [60)], suggesting once again (cf. classic phenol hosts, Fig. 4) that hydrogen bonding plays a fundamental but also a diversified role in clathrate formation [61, 62)].

In 1979, Bishop discovered that a certain rigid *bicyclic aliphatic diol 20* provides a new H-bonded channel inclusion network [63)] (Fig. 9). He expanded his studies to a whole series of compounds similar to the prototype (e.g. *21*) [64, 65)]. As the results show, a suitable interplay of structural features is involved in coming out with an efficient host molecule related to the general building elements. For instance, both

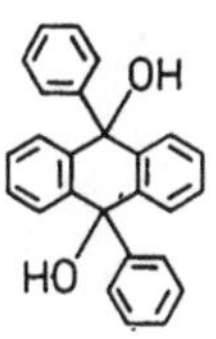

18

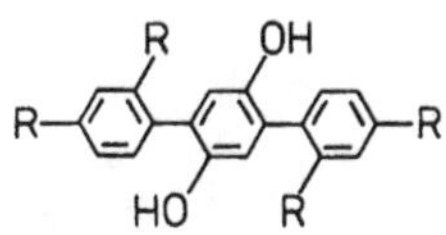

19

R = H , CH_3 , Cl

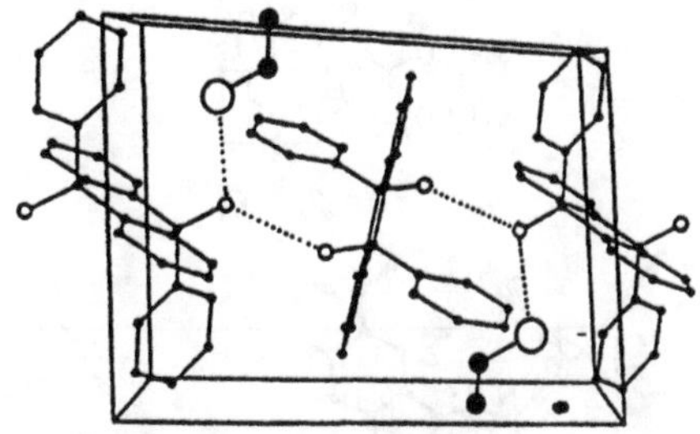

Fig. 8. Examples typical for a new type of hydroxyl group-containing hosts and molecular packing of a corresponding inclusion compound (*18* with ethanol, 1:1; hydrogen bonds as dotted lines; the guest molecules are represented by bold circles and dots) [62]

20

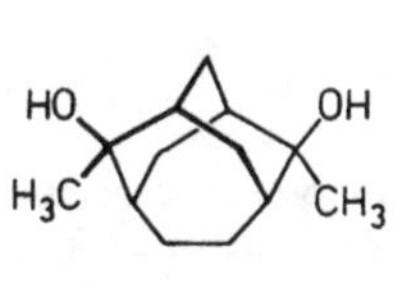

21

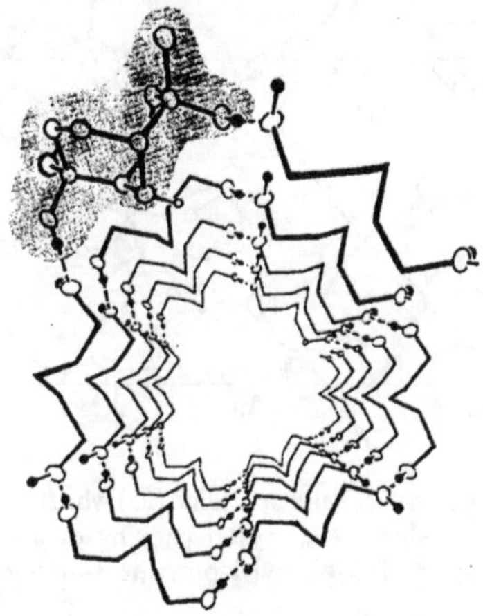

Fig. 9. Constitutions of rigid bicyclic aliphatic diols (*20*, *21*) and diagrammatic representation of the hydrogen bonded helical network developed of *20* (ethyl acetate clathrate, view into an individual channel; guest molecule omitted because of high disorder, one of the host molecules is specified and indicated by shading) [63]

alcohol functions must be tertiary with an α-methyl group, the molecule must have a C_2 rotation axis as the only molecular symmetry, and some further structural restrictions which were specified. Largely abstracted, the main principle of this particular host design combines a rigid skeleton with two appropriately *syn*-positioned hydroxyls which stabilize a hydrogen-bonded channel matrix.

Toda has used *alkaloids* [brucine (*22*), sparteine (*23*)] to enantioselectively enclathrate chiral propargylic alcohols [66,67] (Fig. 10). Hydrogen bonding is an essential part of the lattice build-up. In view of the large pool of natural alkaloids, the potential of the method is obvious.

More recently, Vögtle has discovered due to a chance observation [68], that a certain type of *organic onium compounds* with limited molecular flexibility, e.g. *24* and *25*, is a rich source of clathrate formers [69] (Fig. 11). In addition to the general features of many clathrate hosts, such as bulkiness and limited conformational flexibility, the stability of the ionic host lattice (Fig. 11 b) is certainly of significance for this particular

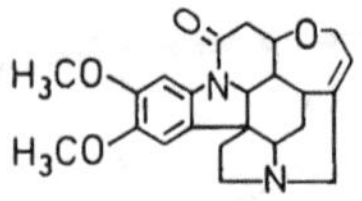

22

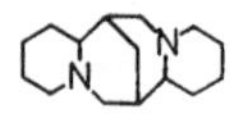

23

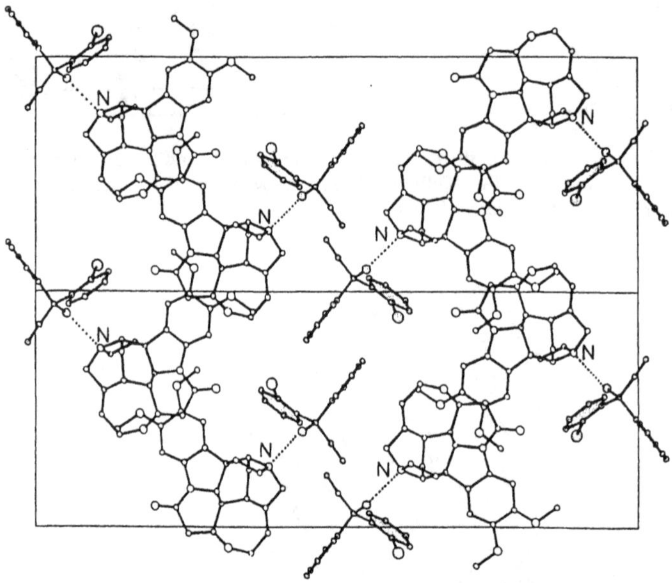

Fig. 10. Constitutions of brucine (*22*) and sparteine (*23*) which are useful for enantioselective inclusion crystallization. The diagram shows the crystal packing of a corresponding clathrate of *22* [guest molecule being 3-(2-bromophenyl)-3-phenyl-1-propyne-3-ol, and is shaded] [66]

clathrate design. The strategy of using ionic forces to create a non-close-packed crystal lattice has also been extended to chiral hosts [70], e.g., quaternized alkaloid bases which allow enantioselective recognition at inclusion formation [71]. But as before, chemoselective guest inclusion does not occur distinctly here.

To overcome this restriction, it has been decided to develop a completely new strategy which is in the formation of "*coordinatoclathrates*" [72] (Fig. 12). The new principle deals with the concerted action of van der Waals non-polar sterical shielding and polar Coulomb attractions or hydrogen bonding. A typical host design involves two main building elements (Fig. 12a): (1) a bulky basic skeleton providing the lattice voids characteristic of a clathrate matrix, and (2) appended function groups (sensor groups) which take the active part in the coordinative host-guest interaction.

This inclusion strategy seems to be the most universal up to now since it is independent on an overall molecular shape, e.g. as characteristic of the above mentioned hosts, and in addition the geometrical as well as the chemical host-guest fit may be controlled by selection of an appropriate building element (bulky skeleton and sensor group). Beside selectivity, coordination-assisted host-guest binding is also suggested to be stronger which will lead to more stable inclusion compounds than under van der Waals conditions alone.

First examples of host constitutions, following the principle of "coordinato-

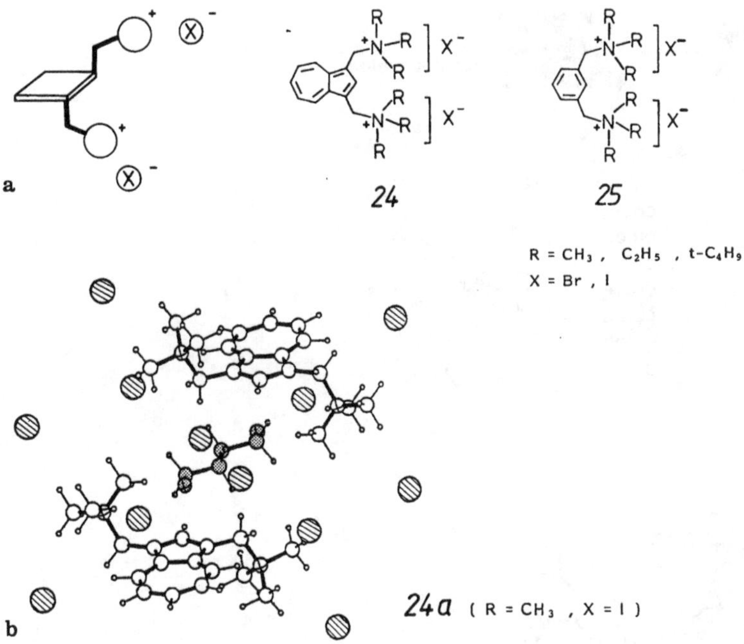

Fig. 11. Organic onium-type clathrate hosts: (**a**) graphic representation of the overall molecular construction; *24* and *25* showing typical host constitutions; (**b**) packing diagram of a representative inclusion compound (*24* with 1-butanol, 1:1; the guest molecules exhibit twofold disorder, and are shaded; iodide ions are hatched) [69]

clathrate formation", confirm the above made assumptions [72–75]. Among others (see volume 2 of this topic), hosts of scissor-like and roof-shaped appearance, e.g. *26* and *27*, having attached carboxylic groups (cf. *26a*), are the most impressive in respect to the formation of selective and stable crystal inclusions (Fig. 12b).

Evidently, the very recent past has made us a present of different profitable design strategies for new clathrate hosts and lattice inclusions. Individual design elements involve molecular bulkiness, rigidity, hydrogen bonding, Coulomb attraction as well as the use of distinguished geometrical bodies, among them those featuring a "wheel-and-axle", a "roof", a "scissor" or a "spider" (see above). Symmetry relations have been noticed by different research groups as a further support in directed clathrate formation [53, 65, 72, 76] (e.g. as concerned with the presence of C_2 or C_3 rotation axis) but there are still more potentialities which are not fully developed.

Nevertheless, rummaging about in the older literature might also be a promising doorway to rediscover "new" clathrate formers (e.g. o-tetraphenylene [76]) and developmental possibilities of known lattice hosts [cf. tri-o-thymotide (*7*) [77], trimesic acid [78], Heilbron "complexes" [79], etc.] are full of surprises.

Summing up, we are now in a fair position to find or to design in future an appropriate host for the great majority of problems being under consideration. The following chapters of this book are giving a more detailed account on the update level of clathrate chemistry with a particular emphasis of the raised questions including selective enclathration, optical resolutions based on clathrate formation, new host

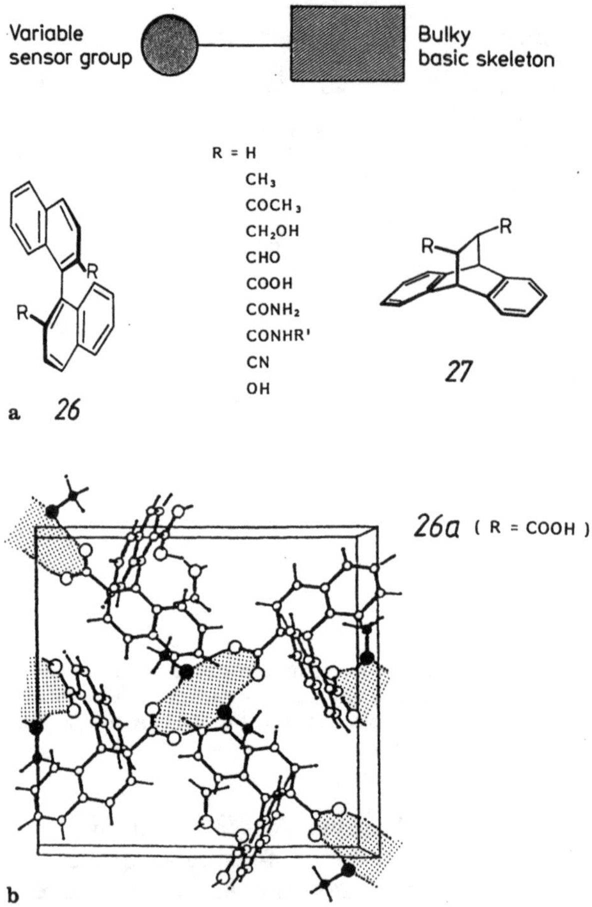

Fig. 12. Coordinatoclathrate concept: **(a)** abstracted structure of a coordinatoclathrate host; *26* and *27* showing prominent examples of molecules (*26* scissor-like, *27* roof-shaped hosts); **(b)** molecular packing of a typical coordinatoclathrate (*26a* with methanol, 1:2; coordinatoclathrate interaction is indicated by the shaded pseudo rings, 12-membered; H-bonds in dotted lines, shaded circles belong to the guest molecules) [72]

constitutions, and structural refinements of lattice inclusions, about in this order. However, before entering into the respective chapters, we shall clear up the terminology to be used (Sect. 4). The reader will profit of this in the later chapters.

4 Calling a Spade a Spade: Classification and Nomenclature of Clathrates (Including Host-Guest Compounds in General)

The term "*clathrate*" was introduced by Powell in 1948 [80] to describe a particular form of molecular compounds in which one component — strictly speaking the host — forms a cage structure imprisoning the other — specified as the guest. It is borrowed from the Latin word "clathratus" (as "closed or protected by cross bars or trellis").

Over the years the original strict sense of the term "clathrate" has become softened up since it was not always used correctly (see ref. 12, pp. 13). The immense increase of new host molecular structures starting off with the discovery of crown compounds [15] in the middle of the sixties has made the situation more difficult [17]. The conventional classification system of inclusion compounds proved to be no longer suitable, because the new compounds can not be conveniently fitted into the existing terminology [81].

For a time one managed by choosing new descriptions and used intricate contractions and modifications of the available terms [82]. Some examples being [83]: clathrate complex, clathrate hydrate, hydrocarbon clathrate, gas hydrate, interlamellar sorbent, molecular compound, addition compound, loose addition complex, cascade complex, lock-and-key complex, super molecular complex, molecular complex associate, tweezer molecule complex, soccer molecule complex, hexapus molecule complex, octopus molecule complex as well as complexes or inclusion compounds of spherands, sepulchrands, coronands, cyclidenes, cryptands, cryptophanes, calixarenes, cucurbituril, annelides etc.

Actually the situation changed for the worse since these descriptions, although pictural in some cases [84], have not been precisely defined, e.g. for a specified inclusion type, or apply only to a very special compound [85]. Heretically speaking, individual authors have attached great importance to use their own descriptions [86] at the risk of increasing the confusion in the literature over the terminology of inclusion compounds of which clathrates are part of, rather than to find a general line [87]. For this very reason a new *system of classification* and *naming* which should be applicable to the presently known and future possible types of inclusion compounds or — what is more — host-guest compounds has been proposed recently [88]. It is now accepted,

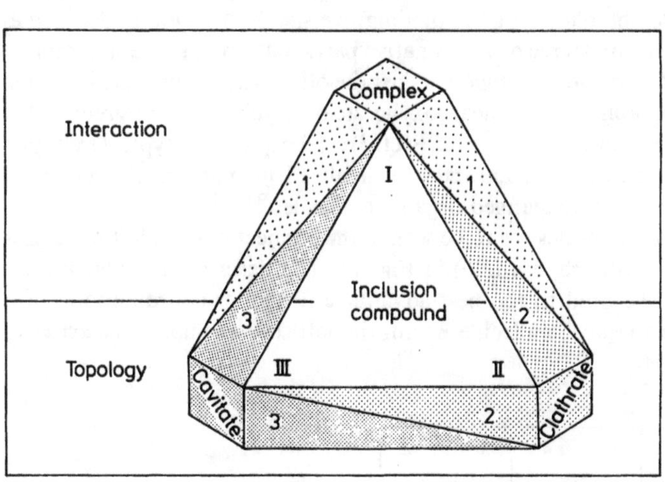

Fig. 13. Classification/nomenclature of host-guest-type compounds; definitions and relations: (1) coordinative interaction, (2) lattice barrier interaction, (3) mono-molecular shielding interaction; (I) coordination-type inclusion compound (inclusion complex), (II) lattice-type inclusion compound (multi-molecular inclusion compound), (III) cavitate-type inclusion compound (mono-molecular inclusion compound)

as literature shows [89], and we also refer to it in this book. Hence a brief illustration is given below.

Two main criteria are essential for advancing the new classification system: (a) the host-guest *interaction* (host-guest type) and (b) the *topology* of the host-guest aggregate (Fig. 13). Regarding (a), a division is made into the extremes of complex and clathrate (or cavitate). Those aggregates which are derived from a *coordination* between host and guest components will be defined as *complexes*. Typical complexes are the metal ion complexes of crown ethers and cryptands [17]. On the other hand, the term "*clathrate*" is reserved to host-guest aggregates where the guest is retained by steric barriers formed by the host lattice (*crystal lattice forces*). Typical examples of this host-guest situation (lattice inclusion) are the inclusion compounds formed by Dianin's compound (*3*), by urea (*1a*), and by graphite [6-12] (see Sect. 1 and Fig. 4). Another approach to make a distinction between complexes and clathrates is to contrast their behaviours in solution: complexes retain their identity in solution, whereas clathrates normally decompose on dissolution.

In respect to (b) (topological aspects) we distinguish between *intra*-molecular host-guest aggregates (*mono*-molecular inclusion compounds or *cavitates* [90]) which operate via any sort of host cavity and *extra*-molecular host-guest aggregates (*multi*-molecular inclusion compounds, equate with *clathrates*) which operate via a lattice void (see above). The same distinction is realized on dissolution: cavitates have an inclination for keeping together in solution (at least if an appropriate solvent is used), but clathrates, as mentioned above, readily decompose. Typical examples of cavitates are the intramolecular cyclodextrin inclusion compounds [6,91,92] (extramolecular lattice inclusion compounds, clathrates, do also exist of this host type).

In many of the recently known host-guest situations (see above), however, both coordinative *and* crystal lattice forces contribute to the formation of the host-guest compound. Thus, borderline cases must be treated as complex/clathrate hybrids (Fig. 14). According to the nature of the binding, we speak of *coordinatoclathrates* which demonstrate a certain degree of coordinative participation but have a dominant clathrate character and of *clathratocomplexes*, just the other way round. The inclusion compounds of 1,1'-binaphthyl-2,2'-dicarboxylic acid (*26a*) with different OH-, NH- and CH-acidic guest molecules [72,75] (cf. Sect. 3) can be quoted as typical examples of coordinatoclathrates; clathratocomplexes are realized in many host-guest compounds of crown ethers with unchanged organic molecules [93-95].

Analogous hybridization modes are to be seen in the connections of all three angles (specified as complex, clathrate, cavitate) in Fig. 13. The description *addition compound* (*adduct*) [96] may be used to the best advantage if neither coordination nor a cavity exist, neither at the host molecule nor in the lattice build-up. *Inclusion com-*

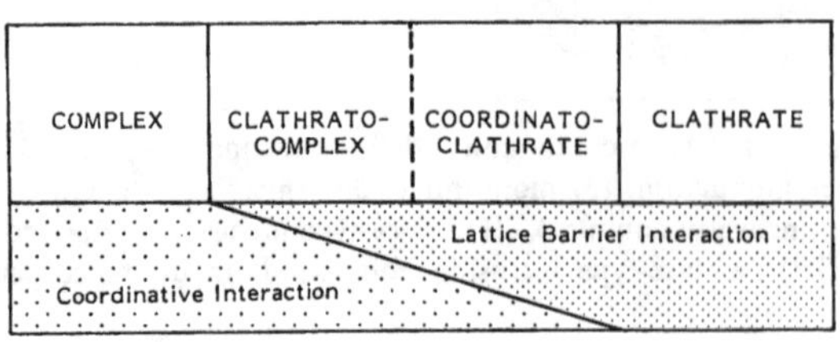

Fig. 14. Assignment of complex/clathrate hybrids

pound[97] is the generic term of choice which refers to the presence of any not precisely defined cavity, whereas the term *host-guest compound*[98] (highest generic term) only expresses a host-guest relationship, whatever. Figure 13 summarizes and interrelates the different terms.

A more detailed *topological* characterization of inclusion compounds according to the shape of the present void (host-type) and its dimension is given in the literature[88]. For instance we have *intercalates*[20] (two-dimensional open, standing for layer- or sandwich-type inclusion compounds), *tubulates* (one-dimensional open, channel inclusions), *aediculates* (applying for pockets or niches), and *cryptates*[16] (totally-enclosed cage inclusions); the terms *coronate* and *podate*[99] apply for ring-shaped and open-chain hosts, respectively.

Numerical considerations dealing with the total number of individual components in a chemical sense, e.g. *binary* (*b*), *ternary* (*t*), etc. and with the number of particles [host, guest separately, e.g. *monomolecular* (*1m*), *binuclear* (*2n*), respectively] are also practicable[88]. Thus, the clathrate formed between Dianin's compound (*3*) and chloroform[100] (see Fig. 4) is identified as binary, hexamolecular, and mononuclear. The full description of this inclusion compound applying the complete set of symbols and designations explained above, hence follows as "b, 6m, 1n-cryptato-clathrate".

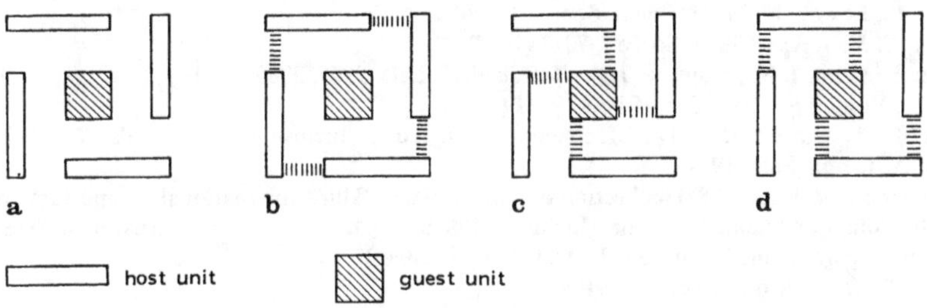

Fig. 15. Diagrammatic representation of different lattice inclusions. Specification (**a–d**) according to host/host and/or host/guest interactions (interactions, e.g. hydrogen bonds, are indicated by broken lines

Indeed, using the proposed nomenclature affords an illustrative and realistic picture of the geometry and interaction within the host-guest unit. But keep in mind, it requires knowledge of structural details, at best of the crystal structure. If so, a very precise characterization via name is possible, e.g. going into detailed H-bonding networks of crystal inclusions (Fig. 15) which are:

(a) *without* any coordinative interaction ("true" clathrate)
(b) coordinative *host-host* interaction (coordination-assisted clathrate host lattice)
(c) coordinative *host-guest* interaction only (coordinatoclathrate)
(d) both coordinative *host-host* and *host-guest* interaction (coordinatoclathrate in a coordination-assisted host lattice).

Illustrations of compounds are found in the following sections.

5 Acknowledgement

The author thanks Dipl. Chem. M. Hecker for drawing the figures.

6 References

1. Davies, J. E. D., Kemula, W., Powell, H. M., Smith, N. O.: J. Incl. Phenom. *1*, 3 (1983)
2. Davy, H.: Philos. Trans. Roy. Soc. (London) *101*, 30 (1811); Faraday, M.: Quart. J. Sci. *15*, 71 (1823)
3. Brown Jr., J. F.: Sci. Am. *207*, 82 (1962)
4. Cramer, F.: Angew. Chem. *64*, 437 (1952)
5. Mandelcorn, L.: Chem. Rev. *59*, 827 (1959)
6. Saenger, W.: Umschau *74*, 635 (1974)
7. Frank, S. G.: J. Pharm. Sci. *64*, 1585 (1975)
8. Davies, J. E. D.: J. Chem. Educ. *54*, 563 (1977)
9. Hagan, M. (ed.): Clathrate Inclusion Compounds, New York, Reinhold Publ. Corp. 1962
10. Mandelcorn, L. (ed.): Non-stoichiometric Compounds, New York, London, Academic Press 1964
11. Bhatnagar, V. M. (ed.): Clathrate Compounds, New Delhi, S. Chand & Co. 1968
12. Gawalek, G. (ed.): Einschlußverbindungen, Additionsverbindungen, Clathrate, Berlin, VEB Deutscher Verlag der Wissenschaft 1969
13. J. Incl. Phenom. *3*, 213 pp (1985) (issue dedicated to H. M. Powell); see also Powell, H. M. in: vol. 1 of ref. 18, p. 1
14. Palin, D. E., Powell, H. M.: J. Chem. Soc. *1947*, 208
15. Pedersen, C. J.: J. Am. Chem. Soc. *89*, 7017 (1967)
16. Dietrich, B., Lehn, J. M., Sauvage, J. P.: Tetrahedron Lett. *1969*, 2885
17. Weber, E., Vögtle, F.: Top. Curr. Chem. *98*, 1 (1981)
18. Atwood, J. L., Davies, J. E. D., MacNicol, D. D. (eds.): Inclusion Compounds, Vol. 1–3, London, Academic Press 1984
19. J. Incl. Phenom. *2*, 1 pp. (1984) (collection of papers of the "Third International Symposium on Clathrate Compounds and Molecular Inclusion Phenomena, The Second International Symposium on Cyclodextrins", July 23–27, 1984, Tokyo, Japan)
20. Davies, J. E. D.: J. Mol. Struc. *75*, 1 (1981)
21. Arad-Yellin, R., Green, B. S., Knossow, M., Tsoucaris, G. in: vol. 3 of ref. 18, p. 263
22. Pollmer, K.: Z. Chem. *19*, 81 (1979)
23. Fetterly, L. C. in: ref. 10, pp. 491
24. Hanotier, J.: Ind. Chim. Belg. *31*, 19 (1966)
25. Fuller, E. J.: Enclyp. Chem. Proc. Des. *8*, 333 (1979)
26. Atwood, J. L. in: vol. 1 of ref. 18, p. 375
27. Sybilska, D., Smolková-Keulemansová, E. in: vol. 3 of ref. 18, p. 173
28. Schmidt, G. M. J., et al.: Solid State Photochemistry, Monographs in Modern Chemistry, Vol. 8, Ginsburg, D. (ed.), Weinheim, New York, Verlag Chemie 1976
29. Gavezzotti, A., Simonetta, M.: Chem. Rev. *82*, 1 (1982)
30. Brown, J. F., White, D. M.: J. Am. Chem. Soc. *82*, 5671 (1960); White, D. M.: J. Am. Chem. Soc. *82*, 5678 (1960)
31. Farina, M., Audisio, G., Natta, G.: J. Am. Chem. Soc. *89*, 5071 (1967); Farina, M., Di Silvestro, G.: J. Chem. Soc., Chem. Commun. *1976*, 816
32. Farina, M. in: vol. 3 of ref. 18, p. 297
33. Dewar, M. J. S., Nahlovsky, B. D.: J. Am. Chem. Soc. *96*, 460 (1974)
34. Arad-Yellin, R., Brunie, S., Green, B. S., Knossov, M., Tsoucaris, G.: J. Am. Chem. Soc. *101*, 7529 (1979)
35. Friedman, N., Lahav, M., Leiserowitz, L., Popovitz-Biro, R., Tang, C. P., Zaretzkii, Z. V. I.: J. Chem. Soc., Chem. Commun. *1975*, 864
36. Setton, R., Beguin, F., Piroelle, S.: Syn. Met. *4*, 299 (1982)
37. Thomas, J. M. in: Intercalation Chemistry, Whittingham, M. S., Jacobsen, A. J. (eds.), New York, Academic Press 1982, p. 55

38. Birell, G. B., Lai, A. A., Griffith, O. H.: J. Chem. Phys. *54*, 1630 (1971)
39. Iwamoto, T., Kiyoki, M., Matsuro, N.: Bull. Chem. Soc. Jpn. *51*, 390 (1978)
40. Lahav, M., Leiserowitz, L., Roitman, L., Tang, C. P.: J. Chem. Soc., Chem. Commun. *1977*, 928
41. MacNicol, D. D., Murphy, A.: Tetrahedron Lett. *22*, 1131 (1981)
42. Garg, S. K., Davidson, D. W., Gough, S. R., Ripmeester, J. A.: Can. J. Chem. *57*, 635 (1979)
43. Parsonage, N. G., Staveley, L. A. K. in: vol. 3 of ref. 18, p. 1
44. Frömming, K. H.: Pharm. uns. Zeit *2*, 109 (1973)
45. Cross, R. J., McKendrick, J. J., MacNicol, D. D.: Nature *245*, 146 (1973)
46. See: Syn. Met. *4*, Nos. 3 and 4 (1982) (special issues devoted to technological applications of synthetic metals)
47. Davies, J. E. D. in: vol. 3 in ref. 18, p. 37
48. Davidson, D. W., Ripmeester, J. A. in: vol. 3 of ref. 18, p. 69
49. Thomas, J. M.: Ultramicroscopy *8*, 13 (1982)
50. Andreetti, G. D. in: ref. 3 of ref. 18, p. 129
51. Powell, H. M. in: ref. 10, p. 438
52. MacNicol, D. D. in: vol. 2 of ref. 18, p. 1
53. MacNicol, D. D., McKendrick, J. J., Wilson, D. R.: Chem. Soc. Rev. *7*, 65 (1978)
54. MacNicol, D. D., Wilson, D. R.: J. Chem. Soc., Chem. Commun. *1976*, 494
55. MacNicol, D. D. in: vol. 2 of ref. 18, p. 123
56. Toda, F., Akagi, K.: Tetrahedron Lett. *1968*, 3695
57. Hart, H., Lin, L. T. W., Ward, D. L.: J. Am. Chem. Soc. *106*, 4043 (1984)
58. Goldberg, I., Lin, L. T. W., Hart, H.: J. Incl. Phenom. *2*, 377 (1984)
59. Hart, H., Lin, L. T. W., Ward, D. L.: J. Chem. Soc., Chem. Commun. *1985*, 293
60. Toda, F., Tanaka, K., Ulibarri Daumas, G., Sanchez, C.: Chem. Lett. *1983*, 1521
61. Toda, F., Tanaka, K., Mak, T.: Chem. Lett. *1983*, 1699
62. Toda, F., Tanaka, K., Mak, T.: Tetrahedron Lett. *25*, 1359 (1984)
63. Bishop, R., Dance, I. G.: J. Chem. Soc., Chem. Commun. *1979*, 992
64. Bishop, R., Choudhury, S., Dance, I. G.: J. Chem. Soc., Perkin Trans. 2, *1982*, 1159
65. Bishop, R., Dance, I. G., Hawkins, S. C.: J. Chem. Soc., Chem. Commun. *1983*, 889
66. Toda, F., Tanaka, K., Ueda, H.: Tetrahedron Lett. *22*, 4669 (1981)
67. Toda, F., Tanaka, K., Ueda, H., Oshima, T.: J. Chem. Soc., Chem. Commun. *1983*, 743
68. Vögtle, F., Löhr, H. G., Puff, H., Schuh, W.: Angew. Chem. *95*, 424 (1983); Angew. Chem., Int. Ed. Engl. *22*, 409 (1983); Angew. Chem. Suppl. *1983*, 527
69. Vögtle, F., Löhr, H. G., Franke, J., Worsch, D.: Angew. Chem. *97*, 721 (1985); Angew. Chem., Int. Ed. Engl. *24*, 727 (1985)
70. Weber, E., Müller, U., Worsch, D., Vögtle, F., Will, G., Kirfel, A.: J. Chem. Soc., Chem. Commun. *1985*, 1578
71. Worsch, D., Vögtle, F., Kirfel, A., Will, G.: Naturwissenschaften *71*, 423 (1984)
72. Weber, E., Csöregh, I., Stensland, B., Czugler, M.: J. Am. Chem. Soc. *106*, 3297 (1984)
73. Czugler, M., Stezowski, J. J., Weber, E.: J. Chem. Soc., Chem. Commun. *1983*, 154
74. Czugler, M., Weber, E., Ahrendt, J.: J. Chem. Soc., Chem. Commun. *1984*, 1632
75. Csöregh, I., Sjögren, A., Czugler, M., Cserzö, M., Weber, E.: J. Chem. Soc., Perkin Trans. 2, *1986*, 507
76. Zheng Huang, N., Mak, T. C. W.: J. Chem. Soc., Chem. Commun. *1982*, 543
77. Ollis, W. D., Stoddart, J. F. in: vol. 2 of ref. 18, p. 169
78. Davies, J. E. D., Finocchiaro, P., Herbstein, F. H. in: vol. 2 of ref. 18, p. 408
79. Herbstein, F. H., Kapon, M., Reisner, G. M., Rubin, M. B.: J. Incl. Phenom. *1*, 233 (1984)
80. Powell, H. M.: J. Chem. Soc. *1948*, 61
81. Problems in naming the different types of inclusion compounds inclusive of clathrates have been recognized very early: Powell, H. M. in: ref. 10, p. 438; see also ref. 9 and ref. 12
82. Cf. Newkome, G. R., Taylor, H. C. R., Fronczek, F. R., Delord, T. J., Kohli, D. K., Vögtle, F.: J. Am. Chem. Soc. *103*, 7376 (1981)
83. Cf. Nickon, A., Silversmith, E. F. (eds.): Organic Chemistry — The Name Game. Modern Coined Terms and Their Origins, New York, Pergamon 1987 (in press)
84. E.g. Vögtle, F., Weber, E.: Angew. Chem. *86*, 896 (1974); Angew. Chem., Int. Ed. Engl. *13*, 814 (1974); Chen, C. W., Whitlock, H. W.: J. Am. Chem. Soc. *100*, 4921 (1978); Gutsche, C. D.: Top. Curr. Chem. *123*, 1 (1984)

85. Mock, W. L., Shih, N. Y.: J. Org. Chem. *48*, 3618 (1983)
86. To be understood as randomly quoted examples: Canceill, J., Collet, A., Gabard, J., Kotzyba-Hibert, F., Lehn, J. M.: Helv. Chim. Acta *65*, 1894 (1982); Canceill, J., Lacombe, L., Collet, A.: J. Am. Chem. Soc. *107*, 6993 (1985)
87. For an early attempt to develop a new classification system for inclusion compounds, see Baron, M.: Physical Methods in Chemical Analysis, Berl, W. (ed.), New York, Academic Press 1961, pp. 259
88. Weber, E., Josel, H.-P.: J. Incl. Phenom. *1*, 79 (1983)
89. E.g. Ungaro, R., Pochini, A., Andreetti, G. D., Domiano, P.: J. Chem. Soc., Perkin Trans. 2, *1985*, 197
90. Cram, D. J.: Science *219*, 1177 (1983)
91. Saenger, W.: Angew. Chem. *92*, 343 (1980); Angew. Chem., Int. Ed. Engl. *19*, 344 (1980); Saenger, W. in: vol. 2 of ref. 18, p. 231
92. Szejtli, J. (ed.): Cyclodextrins and their Inclusion Complexes, Budapest, Akadémiai Kiadó 1982
93. Vögtle, F., Müller, W. M., Watson, W. H.: Top. Curr. Chem. *125*, 131 (1984)
94. Goldberg, I. in: vol. 2 of ref. 18, p. 261
95. Weber, E. in: Progress in Macrocyclic Chemistry, vol. 3, Izatt, R. M., Christensen, J. J. (eds.), New York, Wiley 1986 (in press)
96. This term was originally reserved to designate host-guest compounds which did not appear to be held together by classical chemical bonds and whose bonding was not well understood, cf. ref. 23
97. For an early definition, see Senti, F. R., Erlander, S. R. in: ref. 10, p. 568
98. Cram, D. J., Cram, J. M.: Science *183*, 803 (1974)
99. Weber, E., Vögtle, F.: Inorg. Chim. Acta *L65*, 45 (1980)
100. Flippen, J. L., Karle, J., Karle, I. L.: J. Am. Chem. Soc. *92*, 3749 (1970)

Separation of Enantiomers by Clathrate Formation

Detlev Worsch and Fritz Vögtle

Institut für Organische Chemie und Biochemie der Universität Bonn,
Gerhard-Domagk-Straße 1, D-5300 Bonn 1, FRG

Table of Contents

1 Introduction

Production of enantiomerically pure substances has gained importance, which is far beyond academic interest. The industrial production of large amounts of optically pure compounds today is possible, even if the costs for the comparably laborious procedures may be considerable.

There are two different approaches for the production of enantiomers: asymmetric synthesis and racemate resolution. The first, in special cases, allows a favourable production as costs are concerned, e.g. in case of the asymmetric hydrogenation of double bonds by use of optimized chiral transition metal complexes. In most cases, however, asymmetric syntheses, especially multi-step ones, are restricted to the laboratory scale.

On the other hand, the racemate resolution may be a favourable alternative, if it can be carried out by simple procedures and without loss of substance. That is the reason why racemate resolution or purification of incompletely enriched mixtures of enantiomers by inclusion compounds are so interesting.

Here the ability of chiral host substances to differentiate between guest substances included within their molecular or crystal lattice cavities by diastereomeric interactions is utilized to separate them:

$$\left.\begin{array}{l}(R)\text{-host} + (R)\text{-guest} \rightarrow (R)\text{-host} \cdot (R)\text{-guest} \\[1ex] (R)\text{-host} + (S)\text{-guest} \rightarrow (R)\text{-host} \cdot (S)\text{-guest}\end{array}\right\} \begin{array}{l}\text{diastereomeric} \\ \text{crystal lattices}\end{array}$$

It is essential that *no* covalent bonds are formed between host and guest molecules, as, e.g., when diastereomeric salts are used for separation. The differentiation of the enantiomers is effected only by the "chiral (spatial) environment" in the crystal or in the interior of the host molecule, respectively, whereby hydrogen bridges and dipol-dipol interactions may increase the stability of the inclusion compounds ("coordinato-clathrates" [1]). The different energy constants of the diastereomeric host/guest inclusion compounds which result, e.g., in different solubilities, can be used to isolate one of the enantiomers.

The topic of this survey are the crystal lattice inclusion compounds (clathrates) by use of which racemic substances may be resolved or enriched to yield the enantiomers by simple recrystallization. All methods described here have essentially the following working scheme in common: a) recrystallization of the host compound from the guest solvent or in the presence of the dissolved guest substance from an inert solvent; b) filtration of the clathrate which has formed crystals; c) release of the guest compound by warming or heating (eventually under vacuum). In all cases by repetition of the procedure reusing the enriched guest substance, the enantiomeric excess can be increased, so that usually after three or four recrystallization steps an *ee* (enantiomeric excess) higher than 90% can be obtained.

The simple handling and the special features of this method of separation render the research and development of new host substances an interesting branch of organic stereochemistry. Characteristic features are the following:

a) No or minor loss of substance — the host compound can be recycled.

b) Mild conditions — no high temperature, no strong basic or acidic conditions.
c) Independence of functional groups of host and guest molecules. Guest compounds may even lack functional groups completely.

2 Racemate Resolution by Host Compounds Forming Chiral Crystal Lattices

2.1 Urea

Urea belongs to the clathrate formers which have been used very early for the resolution of racemates [2,3]. Though the enantiomer enrichment usually is not very effective (ca. 5–15 % ee), urea shall still be noted here nevertheless. It shows several peculiarities, which partly are found in other host systems, too. Although the urea molecule is not chiral, it forms chiral lattices**). Regarding their chirality they can be compared to left- and right-handed screws. Chiral long stretched guest molecules can be differentiated inside the ca. 550 pm (5.5 Å) wide channel-type cavities of this lattice. If the crystallization of an enantiomorphic crystal is favoured, only this species will crystallize from the solution and hence can be separated.

Promotion of crystallization can be achieved, e.g., by inoculation with an appropriate enantiomorphic seed crystal, by the presence of further chiral compounds or by use of solvents which already contain a guest enantiomer in small excess. The latter method allows to win the other enantiomer which is now enriched in the filtrate of recrystallization by repetition of the process [2].

The slight difference between the lattice energies of the diastereomeric urea clathrates leads to low ee values and renders the steering of the recrystallization a difficult process which depends strongly on experimental conditions. Therefore, the development of new more universally applicable host substances has largely replaced urea for the resolution of racemates.

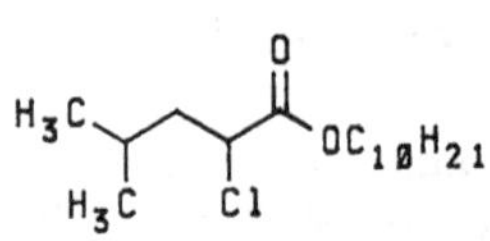

1

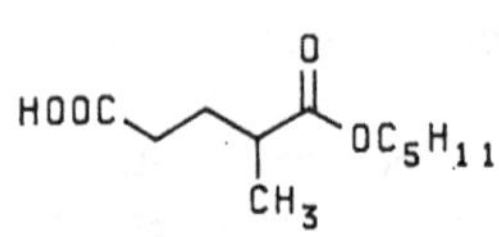

2

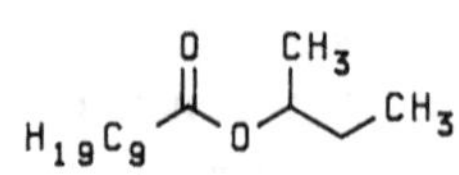

3

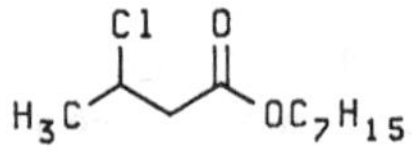

4

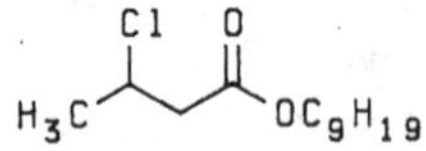

5

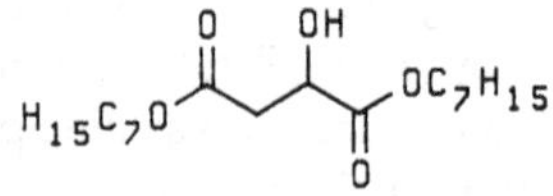

6

---

*) All formulae presented in this paper (except those of α-(*26*) and β-cyclodextrin (*27*) as well as cryptophane (*162*)) were drawn using the "GIOS" computer drawing program, Thieme Verlag, Stuttgart 1985.

**) For details of this general phenomenon see: Addadi, L. Yellin, Z. B. Weissbuch, I., v. Mil, J., Shimon, L. J. W., Lahav, M., Leiserovitz, L.: Angew. Chem. *97*, 476 (1985); Angew. Chem., Int. Ed. Engl. *24*, 466 (1985).

Table 1. Urea as Host Substance for Racemate Resolution

Host molecule	Optical rotation $[\alpha]_D$	Enantiomeric excess ee [%]	Remarks
DL-α-Chloroisocaproic acid decylester (*1*)	−3.18	no remarks	first precipitation
	−5.15	no remarks	second precipitation
	−5.40	no remarks	third precipitation
2-Methylglutaric acid diamylester (*2*) ($[\alpha]_D^{20} = +3.45$)	+8.06	no remarks	—
DL-Capric acid-2-butylester (*3*)	+1.17	no remarks	crystallization upon acetate silk
β-Chlorobutyric acid heptylester (*4*)	−1.03	no remarks	in the presence of (+)-tartaric acid
β-Chlorobutyric acid nonylester (*5*)	+1.83		guest exchange using *n*-heptane-urea clathrate
Malic acid diheptylester (*6*)	+3.8		twofold recrystallization
	+1.23	no remarks	inoculation in the course of the "three chamber procedure" (cf. ref. 2)

2.2 Tri-*o*-thymotide ("TOT")

Tri-*o*-thymotide ("TOT", *7*) has become known as a versatile clathrate former[4−8]. The molecule can exist in two helical chiral propeller-shaped conformers, which at room temperature rapidly convert into each other (ΔG ca. 88 kJ/mol). In the absence of appropriate guest molecules TOT crystallizes in an achiral crystal lattice of the space group $Pna2_1$ which contains molecules of the *P*- (plus) and *M*- (minus) helical configuration. However, if TOT crystallizes under inclusion of guest substances, in other words as a clathrate, spontaneous racemate resolution usually results:

Single clathrate crystals consist of *P*- or *M*-host molecules, respectively (formation of a conglomerate), which because of their chirality have enclosed one of the enantiomers preferentially. This peculiarity of the crystallization behaviour allows the resolution of racemic guest molecules. The crystals of the clathrate are identified by solving a crystal splitter and measuring the optical rotation. (Usually, the optical rotation of TOT outweighs those of the guests because of its high molecular mass.) Crystals containing TOT in the *P*- or *M*-configuration are used separately as seed crystals to grow larger single crystals[6]. The guest compound is easily isolated from these single crystals by warming in vacuum. As TOT, depending on the shape of the guest molecules can form channel as well as cage type inclusions, a broad spectrum of guest substances can be used. Compounds having a constitution analogous to TOT, in which the three oxygen atoms in the macrocyclic ring, for instance, are substituted by nitrogen atoms, also show spontaneous racemate resolution combined with clathrate formation[5]. The properties of these related host substances, as regards differentiation of enantiomers, are only little investigated as yet. These host compounds and the TOT molecule itself possess a three-fold axis of symmetry, which is encountered remarkably often with good clathrate formers: The cyclophosphazenes[9], perhydrotriphenylene[10], hexa hosts[11−13] (as part of the six-fold symmetry axis), Dianins compound[14] (three-fold lattice symmetry).

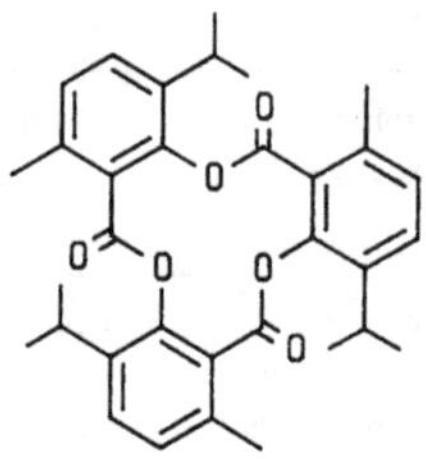

7

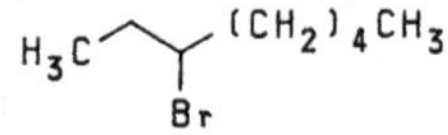

14

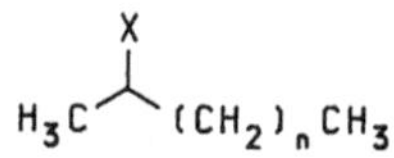

	X	n
8	Cl	1
9	Br	1
10	NH_2	1
11	OH	1
12	Cl	5
13	Br	5
15	Br	6
16	Br	9

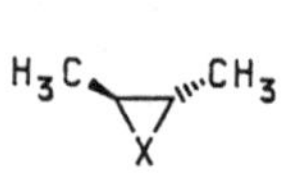

17: X = O
18: X = S

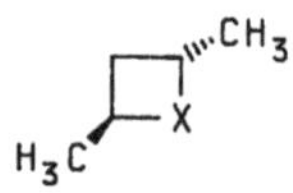

19: X = O
20: X = S

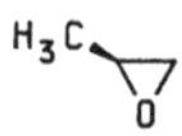

21

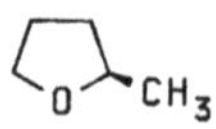

22

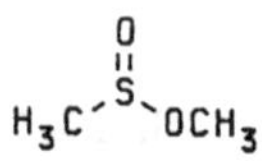

23

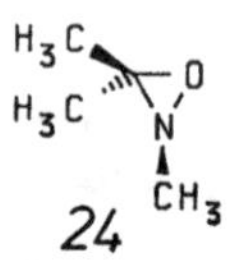

24

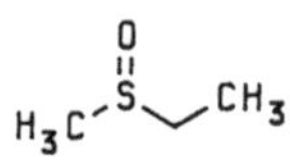

25

Table 2. Resolution of Racemates Using (P)-(+)-Tri-*o*-thymotide (TOT, *7*)[6-8]

Guest molecule	ee [%]	Confi-guration	Type of Clathrate	Remarks
2-Chlorobutane (*8*)	32–45	(*S*)-(+)	cage	
2-Bromobutane (*9*)	34–37	(*S*)-(+)	cage	
2-Butane amine (*10*)	<2	no remarks	no remarks	
2-Butanol (*11*)	<5	no remarks	no remarks	
2-Chlorooctane (*12*)	4	(*S*)-(+)	channel	
2-Bromooctane (*13*)	4	(*S*)-(+)	channel	
3-Bromooctane (*14*)	4	(*S*)-(+)	channel	
2-Bromononane (*15*)	5	(*S*)-(+)	channel	
2-Bromododecane (*16*)	5	(*S*)-(+)	channel	
trans-2,3-Dimethyloxirane (*17*)	47	(*S,S*)-(−)	cage	
trans-2,3-Dimethylthiirane (*18*)	30	(*S,S*)-(−)	cage	
trans-2,4-Dimethyloxetane (*19*)	38	no remarks	cage	
trans-2,4-Dimethylthietane (*20*)	9	no remarks	cage	only one crystal was measured
Propylene oxide (*21*)	5	(*R*)-(+)	cage	
2-Methyltetrahydrofurane (*22*)	2	(*S*)-(+)	cage	
Methylsulfinic acid methylester (*23*)	14–15	(*R*)-(+)	cage	
2,3,3-Trimethyl-oxaziridine (*24*)	7	no remarks	cage	
Ethylmethylsulfoxide (*25*)	40–80	(*R*)-(+)	cage	

2.3 Cyclodextrins

Whereas the separation of racemates in the case of urea and TOT was achieved only by a chiral crystal lattice of the achiral or racemic host, respectively, the optically active cyclodextrins, available from the chiral pool, are able to differentiate a chiral guest within their intramolecular cavity. Therefore, they do not necessarily need the crystal lattice to form inclusion compounds. The guest is encapsulated, while is is in solution, too, if the guest by size and shape fits into the cavity of the specific cyclodextrin molecule (α- (*26*), β- (*27*), or γ-cyclodextrins).

This particular property allows an application of the cyclodextrins as chiral material for chromatographic racemate resolution [15,16] or as substrates for the asymmetric induction leading to bond closure or bond cleavage (enzyme modelling). Apart from the topological (static) conditions, the dynamic processes of complex formation and complex dissociation here also play a part in the host/guest interactions and are therefore important for the efficiency of the separation effect.

The examples of racemate resolution noted below all are based on crystalline

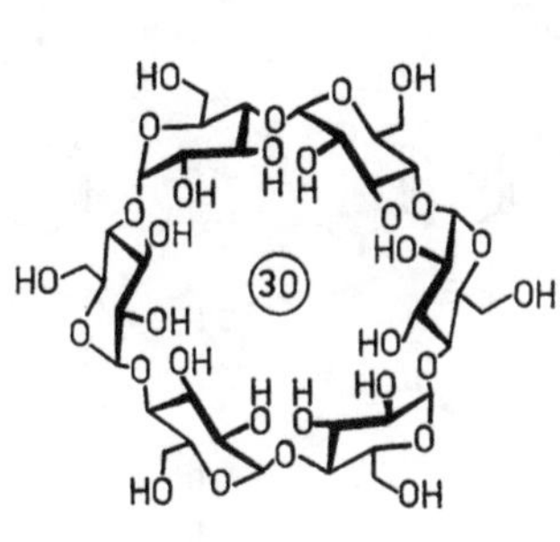

26

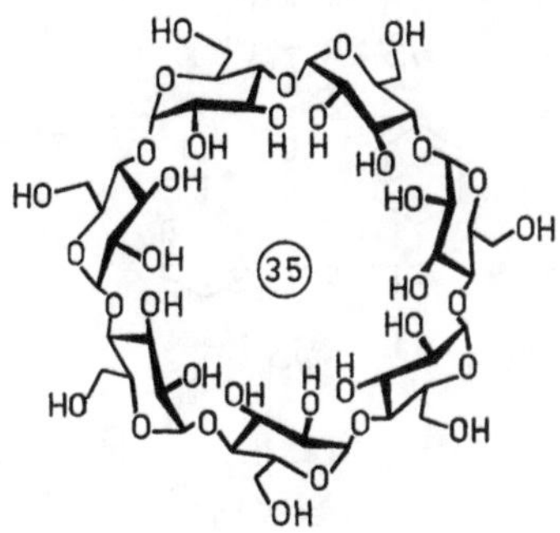

27

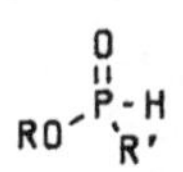

	R	R'
28	C_2H_5	C_2H_5
29	C_2H_5	C_6H_5
30	i-C_3H_7	CH_3
31	i-C_3H_7	C_2H_5

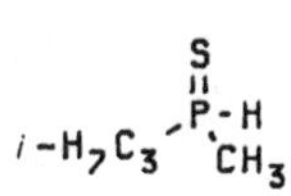

32

$[Co(en)_2(H_2NCH_2CH_2SR)]\ X_3$

	R	X
33	$C_{10}H_{20}Br$	Cl
34	$C_8H_{16}Br$	Cl
35	$C_{10}H_{20}Br$	Br
36	$C_8H_{16}Br$	Br
37	$C_{10}H_{20}COOH$	Br
38	$C_{12}H_{25}$	Br
39	$C_{10}H_{21}$	Br
40	C_8H_{17}	Br

F_3C-CHClBr

41

cyclodextrin clathrates. Here, the spatial lattice structure is in the centre of view. Clathrate formation using α- and β-cyclodextrins (*26* and *27*) allowed an elegant resolution of racemic sulfoxides, sulfinic and phosphonic acid esters which have been hardly available up to now.

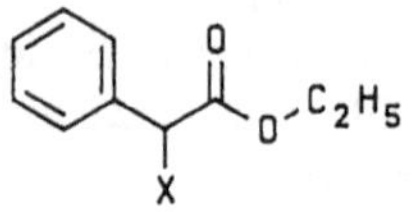

42: X = Cl
43: X = Br

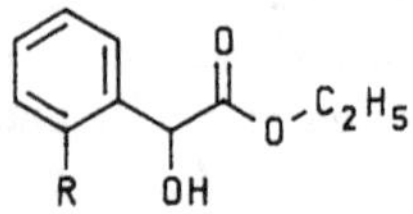

44: R = H
45: R = Cl

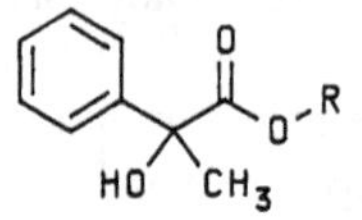

46: R = H
47: R = C_2H_5

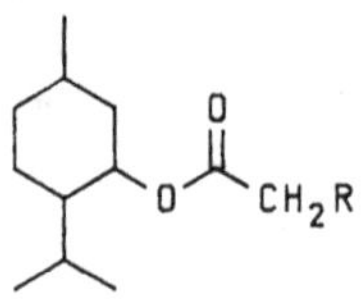

48: R = H
49: R = Cl

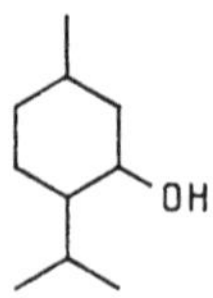

50

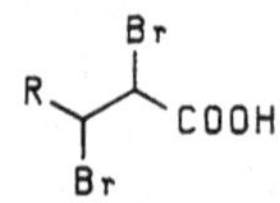

51: R = C_6H_5
52: R = COOH

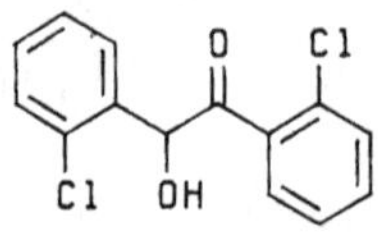

53

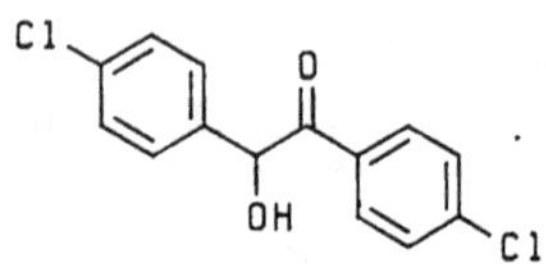

54

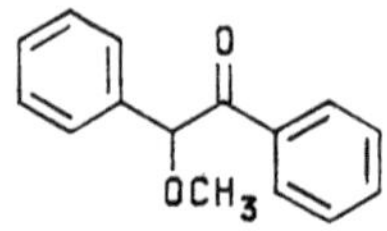

55

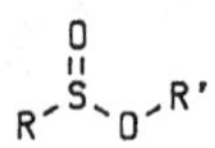

	R	R'
56	CH_3	*n*-C_3H_7
57	CH_3	*i*-C_3H_7
58	CH_3	*i*-C_4H_9
59	CH_3	*tert*-C_4H_9
60	CH_3	cyclopentyl
61	CH_3	$CH_2C(CH_3)_3$
62	*i*-C_3H_7	CH_3
63	*i*-C_3H_7	*i*-C_3H_7

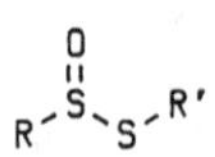

	R	R'
64	*tert*-C_4H_9	*tert*-C_4H_9
65	*tert*-C_4H_9	CH_3
66	*p*-$C_6H_4CH_3$	*tert*-C_4H_9

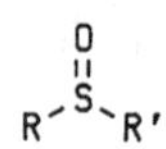

	R	R'
67	$CH_2C_6H_5$	CH_3
68	$CH_2C_6H_5$	C_2H_5
69	$CH_2C_6H_5$	*n*-C_3H_7
70	$CH_2C_6H_5$	*n*-C_4H_9
71	$CH_2C_6H_5$	*i*-C_4H_9
72	$CH_2C_6H_5$	*tert*-C_4H_9
73	C_6H_5	CH_3
74	C_6H_5	C_2H_5
75	C_6H_5	*i*-C_3H_7
76	C_6H_5	*n*-C_4H_9
77	C_6H_5	*i*-C_4H_9
78	C_6H_5	*tert*-C_4H_9
79	*p*-$C_6H_4CH_3$	CH_3
80	*p*-$C_6H_4CH_3$	C_2H_5
81	*p*-$C_6H_4CH_3$	*i*-C_3H_7
82	*p*-$C_6H_4CH_3$	*n*-C_4H_9
83	*p*-$C_6H_4CH_3$	*tert*-C_4H_9

The optical enrichment of halothan, $CF_3CHBrCl$ (*41*), using α-cyclodextrin (*26*) [17], is the only known strategy hitherto for racemate resolution of this substance. This is a good example to demonstrate the independency of the inclusion method on functional groups, even if the optical yield is only low as yet. These are indispensable when "classical" methods of resolution, e.g. use of diastereomeric salts, are applied; consequently their application field is very restricted.

Table 3. Resolution of Racemates Using α-Cyclodextrin (*26*) [18]

Guest molecule	Optical rotation $[\alpha]_D^{25}$	ee [%]	Stoichiometry host: guest
Ethyl-phosphinic acid ethylester (*28*)	—5.06	23.8	
Phenyl-phosphinic acid ethylester (*29*)	12.08	28.8	
Methylthio phosphinic acid *i*-propylester (*32*)	14.92	no remarks	
Cobalt complex *33*	no remarks	61	2:1
34	no remarks	9.2	1.3:1
35	no remarks	30	1.5:1
36	no remarks	12	0.88:1
37	no remarks	12	2:1
38	no remarks	20	2:1
39	no remarks	6.4	0.7:1
40	no remarks	53	2:1

2.4 Host Molecules Derived from Binaphthyl, and Related Skeletons

The well known chiral carbon skeleton designated as binaphthyl hinge has been introduced into asymmetric synthesis and resolution of racemates in the form of the derivatives of 2,2'-dihydroxy-1,1'-binaphthyl (*84*, binaphthol). The application of chiral crown compounds containing this binaphthyl unit for the separation of amino acids and amino acid esters by use of liquid/liquid chromatography has been described particularly by Cram et al. in detail [23-37].

Crystalline inclusion compounds containing organic sulfoxides as guests are known of binaphthol (*84*) and of the structurally similar spiro compound *92*. These clathrates have been used to separate the enantiomers of those guest compounds [38,39]. Here, another peculiarity of racemate resolution by clathrate formation is shown: The host compound has not necessarily to play the dominant steering role even if it provides the crystal lattice and the guest fits in. By using optically pure binaphthol (*84*), the enantiomers of racemic sulfoxides have been separated; on the other hand, optical enrichment of racemic binaphthols by using racemic pure sulfoxides has also been successfully carried out [39].

The spiro compound *92* and the alkaloid sparteine (*85*), being racemates, analogously have been successfully resolved by using the optically pure partner [38].

Table 4. Resolution of Racemates Using β-Cyclodextrin (*27*)[18–22]

Guest molecule	Optical rotation $[\alpha]_D^{25}$	ee [%]	Remarks
Methyl-phosphinic acid *i*-propylester (*30*)	−17.30	66.5	
	−21.86	84.0	after threefold recrystallization
Ethyl-phosphinic acid *i*-propylester (*31*)	−15.18	60.0	
Cobalt complex *33*	no remarks	2.0	stoichiometry host:guest = 1:1
34	no remarks		
35	no remarks	1.5	
36	no remarks	3.3	
38	no remarks	0.8	stoichiometry host:guest = 1.33:1
39	no remarks	2.9	stoichiometry host:guest = 1:1
Halothan (*41*)	$[\alpha]_{365}^{24.5}$ −0.025	no remarks	first step
	$[\alpha]_{365}^{24.5}$ −0.027	no remarks	second step
	$[\alpha]_{365}^{24.5}$ −0.029	no remarks	third step
Phenylchloroacetic acid ethylester (*42*)	−3.40	3.15	molar ratio host:guest = 1:8
Phenylbromoacetic acid ethylester (*43*)	−0.92	5.75	ethanol as solvent
Mandelic acid ethylester (*44*)	−5.90	3.32	CS_2 as solvent
o-Chloromandelic acid ethylester (*45*)	−12.65	no remarks	molar ratio host:guest = 1:8
2-Hydroxy-2-phenylpropionic acid (*46*)	−2.90	5.57	water as solvent
2-Hydroxy-2-phenylpropionic acid ethylester (*47*)	−2.23	8.35	ethanol as solvent
Acetic acid menthylester (*48*)	−1.42	1.78	methanol as solvent
Monochloroacetic acid menthylester (*49*)	−1.72	2.21	methanol as solvent
Menthol (*50*)	−2.44	4.88	ethanol as solvent
Dibromocinnamic acid (*51*)	+7.71	11.33	ethanol as solvent
Dibromosuccinic acid (*52*)	+12.10	8.18	ethanol as solvent
2,2′-Dichlorobenzoin (*53*)	+1.50	no remarks	ethyl acetate as solvent
4,4′-Dichlorobenzoin (*54*)	+4.62	no remarks	ethyl acetate as solvent
Benzoin methylether (*55*)	+0.79	0.84	ethanol as solvent
Methylsulfinic acid *n*-propylester (*56*)	2.38	1.4	stoichiometry host:guest = 1:1[a]
Methylsulfinic acid *i*-propylester (*57*)	−165.91	70.20	stoichiometry host:guest = 1:1[b]
Methylsulfinic acid *i*-butylester (*58*)	10.10	8.70	stoichiometry host:guest = 2:1[a]
Methylsulfinic acid *tert*-butylester (*59*)	−19.90	12.40	stoichiometry host:guest = 4:3[b]
Methylsulfinic acid cyclopentylester (*60*)	1.10	0.40	stoichiometry host:guest = 1:1[a]

Methylsulfinic acid neopentylester (*61*)	5.16	4.20	stoichiometry host:guest = 1:1[a]
Isopropylsulfinic acid methylester (*62*)	+14.40	4.20	stoichiometry host:guest = 1:1[a]
Isopropylsulfinic acid *i*-propylester (*63*)	−1.87	2.10	stoichiometry host:guest = 1:1[a]
tert-Butylsulfinic acid *tert*-butylthioester (*64*)	−21.10	13.6	[a]
tert-Butylsulfinic acid thiomethylester (*65*)	−3.04	no remarks	no remarks
p-Methylphenylsulfinic acid *tert*-butylthioester (*66*)	2.00	2.5	[a]
Benzylmethylsulfoxide (*67*)	−8.50	8.0	[a]
Benzylethylsulfoxide (*68*)	5.00	4.7	[a]
Benzyl-*n*-propylsulfoxide (*69*)	2.00	3.6	[b]
Benzyl-*n*-butylsulfoxide (*70*)	1.96	2.0	[a]
Benzyl-*i*-butylsulfoxide (*71*)	6.40	5.8	[a]
Benzyl-*tert*-butylsulfoxide (*72*)	45.00	14.5	[a]
Methylphenylsulfoxide (*73*)	6.50	4.4	[a]
Ethylphenylsulfoxide (*74*)	16.10	9.1	[a]
i-Propylphenylsulfoxide (*75*)	8.80	5.2	[a]
n-Butylphenylsulfoxide (*76*)	15.30	9.2	[a]
i-Butylphenylsulfoxide (*77*)	4.90	2.2	[a]
tert-Butylphenylsulfoxide (*78*)	1.90	1.1	[b]
p-Methylphenylmethylsulfoxide (*79*)	11.50	8.0	[a]
p-Methylphenylethylsulfoxide (*80*)	11.10	5.3	[a]
p-Methylphenyl-*i*-propylsulfoxide (*81*)	3.45	1.0	[a]
p-Methylphenyl-*n*-butylsulfoxide (*82*)	6.70	3.6	[a]
p-Methylphenyl-*tert*-butylsulfoxide (*83*)	−12.40	6.4	[b]

[a] Configuration of the guest: (*R*). — [b] Configuration of the guest: (*S*).

Table 5. Resolution of Enantiomers Using (*R*)-(+)-Binaphthol (*84*) [38, 39]

Host molecule	ee [%]	Configu-ration	Remarks
Ethylmethylsulfoxide (*25*)	25	(+)	(−) with (*S*)-(−)-*84*
Methylphenylsulfoxide (*73*)	5		
Methyl-*m*-methylphenylsulfoxide (*86*)	100	(+)	separation by chromato-graphy on silica gel
	62	(−)	from mother liquid
	100	(−)	by application of 62% ee (−)-*86* and (*S*)-(−)-*84*
Ethyl-*m*-methylphenylsulfoxide (*87*)	100	(+)	
	100	(−)	with (*S*)-(−)-*84*
m-Methylphenylvinylsulfoxide (*88*)	100	(+)	cf. ref. [38]
n-Butylmethylsulfoxide (*89*)	100	(+)	(−) with (*S*)-(−)-*84*
Methyl-*n*-propylsulfoxide (*90*)	100	(+)	(−) with (*S*)-(−)-*84*
i-Butylmethylsulfoxide (*91*)	25	(+)	(−) with (*S*)-(−)-*84*
Methyl-*m*-methylphenylsulfoximine (*93*)	100	(+)	
Ethyl-*m*-methylphenylsulfoximine (*94*)	100	(+)	

Reversely, in most cases the enantiomers of binaphthol could be separated completely using the optically pure guest compound.

Table 6. Resolution of Enantiomers Using 2,2′-Dihydroxy-9,9′-spirodifluorene (*92*)

Guest molecule	ee [%]	Remarks
Sparteine (*85*)		cf. ref. [38]
3-Methylpiperidine (*95*)		cf. ref. [38]

The enantiomers of the host compound *92* could be separated starting with the racemate using an optically active guest (sparteine), too.

2.5 Alkaloids

If alkaloids are mentioned in connection with racemate resolution, one is usually inclined to think at first of the classical methods of resolution using diastereomeric salt formation by combination of the alkaloid base and an organic acid.

Though this method has given satisfactory results in many cases, it has the disadvantage of depending on functional groups (acid function). However, beside salt formation in some cases formation of clathrates of alkaloids plays a decisive role in racemate resolution processes. As an example the enantiomeric enrichment of bromochlorofluoromethane (CHClBrF) was achieved successfully by means of *brucine* [40]. Typically enough measurement of the enantiomeric excess (ee) of this interesting anaesthetic was achieved by differentiation of the enantiomers in a cavity of the cyclophane type [41] (cf. section 2.6).

In addition, the alkaloids listed here, *sparteine* (*85*), *brucine* (*96*) and *quinine* (*97*), can be used to differentiate between the enantiomers of racemic guest compounds

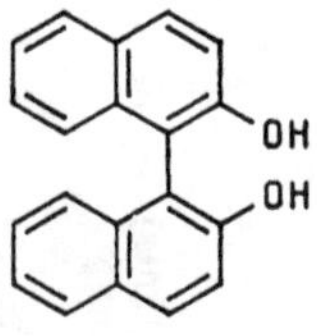

84

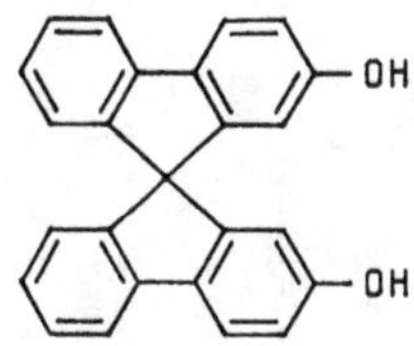

92

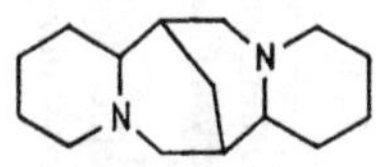

85

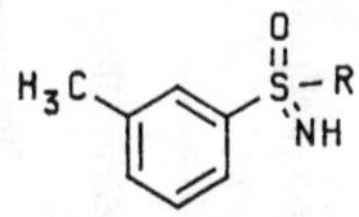

93: R = CH_3
94: R = C_2H_5

	R	R'
86	CH_3	*m*-$C_6H_4CH_3$
87	C_2H_5	*m*-$C_6H_4CH_3$
88	$CH{=}CH_2$	*m*-$C_6H_4CH_3$
89	CH_3	*n*-C_4H_9
90	CH_3	*n*-C_3H_7
91	CH_3	*i*-C_3H_7

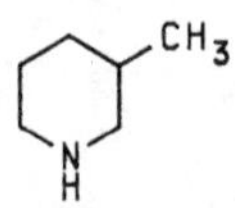

95

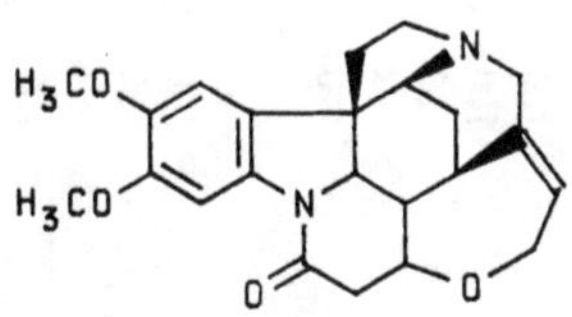

96

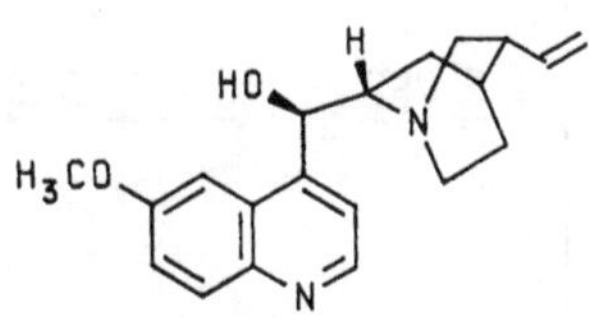

97

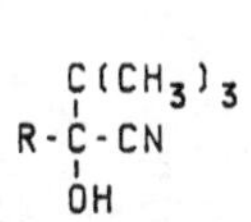

	R
98	C_6H_5
99	*p*-C_6H_4Cl
100	*p*-$C_6H_4CH_3$
101	*p*-C_6H_4OH
102	*m*-C_6H_4OH

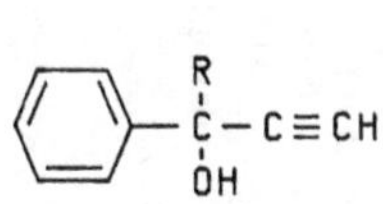

	R		R
104	C_2H_5	111	*o*-C_6H_4Br
105	*tert*-C_4H_9	112	*o*-C_6H_4F
106	*i*-C_3H_7	113	*o*-C_6H_4Cl
107	*tert*-C_5H_{11}	114	CCl_3
108	*n*-C_3H_7	115	$CHCl_2$
109	*n*-C_4H_9	116	CH_2Cl
110	*o*-$C_6H_4CH_3$		

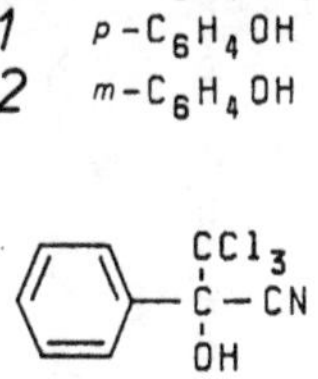

103

Table 7. Resolution of Enantiomers Using Brucine (*96*) [38, 42–45]

Guest molecule	ee [%]	$[\alpha]_D$	Configuration (opt. rotation)	Notes
1-Cyano-2,2-dimethyl-1-phenylpropanol (*98*)	97	+15.5	d	
1-Cyano-2,2-dimethyl-1-*p*-chlorophenylpropanol (*99*)	100	+5.5	d	
1-Cyano-2,2-dimethyl-1-*p*-methylphenylpropanol (*100*)	52	−2.6	l	
	100	−5.0	l	after recrystallization (three times)
1-Cyano-2,2-dimethyl-1-*p*-hydroxyphenylpropanol (*101*)		−7.3	l	
1-Cyano-2,2-dimethyl-1-*m*-hydroxyphenylpropanol (*102*)		−13.0	l	
1-Cyano-2,2,2-trichloro-1-phenylethanol (*103*)	100	8.1	d	
3-Phenyl-1-pentin-3-ol (*104*)	100	7.2	(+)	after recrystallization (2 times)
4,4-Dimethyl-3-phenyl-1-pentin-3-ol (*105*)	100	8.8	(*R*)-(+)	
	93	−8.2	(*S*)-(−)	from mother liquid
4-Methyl-3-phenyl-1-pentin-3-ol (*106*)	100	1.1	(+)	after recrystallization (3 times)
5,5-Dimethyl-3-phenyl-1-hexin-3-ol (*107*)	100	+10.5	(*R*)-(+)	
	70	−7.3	(*S*)-(−)	from mother liquid
3-Phenyl-1-hexin-3-ol (*108*)	100	4.5	(+)	after recrystallization (2 times)
3-Phenyl-1-heptin-3-ol (*109*)	100	7.9	(+)	
3-*o*-Methylphenyl-3-phenyl-1-propin-3-ol (*110*)	68	−15.7	(−)	
	94	21.1	(+)	from mother liquid
	100	−53.7	(−)	after recrystallization (6 times)
3-*o*-Bromophenyl-3-phenyl-1-propin-3-ol (*111*)	46	−52.1	(−)	
	95	108	(+)	from mother liquid
	100	−114	(−)	after recrystallization (3 times)
3-*o*-Fluorophenyl-3-phenyl-1-propin-3-ol (*112*)	60	−35.4	(−)	
	53	31.3	(+)	from mother liquid
	100	−59.6	(−)	after recrystallization (3 times)
3-*o*-Chlorophenyl-3-phenyl-1-propin-3-ol (*113*)	100	−129	(−)	after recrystallization (8 times)
4,4,4-Trichloro-3-phenyl-1-butin-3-ol (*114*)	100	13.8	(+)	after recrystallization (x times)
4,4-Dichloro-3-phenyl-1-butin-3-ol (*115*)	100	−3.4	(−)	after recrystallization (2 times)
4-Chloro-3-phenyl-1-butin-3-ol (*116*)	100	11.1	(+)	after recrystallization (3 times)
3-*o*-Chlorophenyl-3-*m*-methylphenyl-1-propin-3-ol (*117*)	100	−126	(−)	after repeated recrystallization

3-α-Naphthyl-3-phenyl-1-propin-3-ol (*118*)	100	116	(+)	after recrystallization (5 times)
3-α-Naphthyl-1-pentin-3-ol (*119*)	100	55.4	(+)	after recrystallization (2 times)
4-Chloro-3-α-naphthyl-1-butin-3-ol (*120*)	100	58.1	(+)	after recrystallization (3 times)
3-α-Thiopheno-1-octin-3-ol (*121*)	100	5.5	(+)	after recrystallization (2 times)
3-α-Thiopheno-1-heptin-3-ol (*122*)	100	3.1	(+)	after recrystallization (2 times)
3-α-Thiopheno-1-hexin-3-ol (*123*)	100	3.9	(+)	after recrystallization (2 times)
4-Methyl-3-α-thiopheno-1-pentin-3-ol (*124*)	100	4.5	(+)	after recrystallization (2 times)
5-Chloro-3,5-diphenyl-1-pentin-3-ol (*125*)	100	11.6	(+)	after recrystallization (3 times)
5-Bromo-3-phenyl-1-hexin-3-ol (*126*)	100	—3.0	(—)	after recrystallization (3 times)
5-Methoxy-3-phenyl-1-hexin-3-ol (*127*)	100	1.6	(+)	after recrystallization (3 times)
(ratio of diastereomers = 40:60 related to atom 3)				
4-Chloro-3-phenyl-1-pentin-3-ol (*128*)	16	no remarks	(+)	
(ratio of diastereomers 89:11 related to atom 3)	100	6.3	(+)	after recrystallization (4 times)
4-Chloro-3,4-diphenyl-1-butin-3-ol (*129*)	100	23.8	(+)	after recrystallization (4 times)
(ratio of diastereomers 20:80 related to atom 3)				
4-Chloro-3-thiopheno-1-pentin-3-ol (*130*)	100	12.2	(+)	after recrystallization (4 times)
(ratio of diastereomers 87:13 related to atom 3)				
4,5-Dichloro-3-phenyl-1-hexin-3-ol (*131*)	100	39.0	(+)	after recrystallization (3 times)
(ratio of diastereomers 5:3 related to atom 3)				
Bromochlorofluoromethane (*133*)	4.3	0.128	(+)	cf. ref. [40, 41]
2-Cyano-7-oxabicyclo[2.2.1]hept-5-en-2-ylacetate (*137*)	>99	57.9	(+)	cf. ref. [54]

Table 8. Resolution of Racemates Using Sparteine (*85*) [38, 43]

Guest molecule	ee [%]	$[\alpha]_D$	Configuration of the guest	Remarks
2,2′-Dihydroxy-9,9′-spirodifluorene (*92*)	no remarks	no remarks	no remarks	cf. chapter 2.4.
4,4-Dimethyl-3-phenyl-1-pentin-3-ol (*105*)	59	−7.3	(−)	
	57	7.1	(+)	from the mother liquid
3-*o*-Bromophenyl-3-phenyl-1-propin-3-ol (*111*)	50	−66.7	(−)	
	81	109.0	(+)	from the mother liquid
3-*o*-Fluorophenyl-3-phenyl-1-propin-3-ol (*112*)	34	−20.0	(−)	
	23	13.9	(+)	from the mother liquid
3-*o*-Chlorophenyl-3-phenyl-1-propin-3-ol (*113*)	55	−73.7	(−)	
	60	+81.3	(+)	from the mother liquid

A complete resolution of the (—)-enantiomer of racemic sparteine was achieved using optically pure *111* [43].

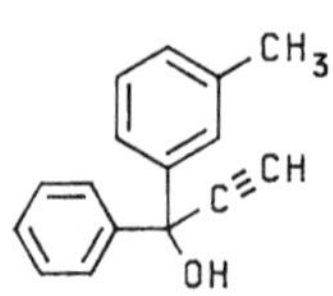

117

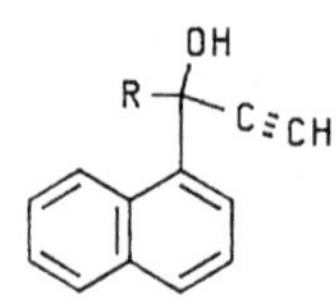

	R
118	C_6H_5
119	C_2H_5
120	CH_2Cl

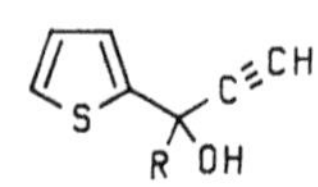

	R
121	n-C_5H_{11}
122	n-C_4H_9
123	n-C_3H_7
124	i-C_3H_7

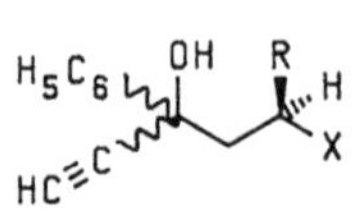

	R	X
125	C_6H_5	Cl
126	CH_3	Br
127	CH_3	OCH_3

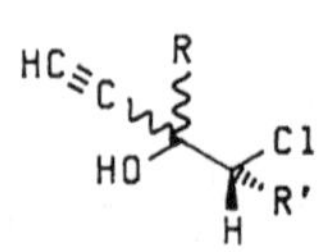

	R	R′
128	C_6H_5	CH_3
129	C_6H_5	C_6H_5
130	thiophene	CH_3

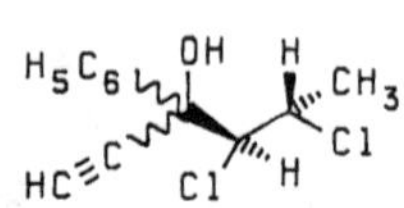

131

by clathrate formation, either as free bases [38, 40, 42–45] or as quaternary onium salts [46].

The remarkable inclusion capacity of organic onium salts has been shown for numerous representatives of this family of host compounds [47–53].

Thus it is obvious that optically pure onium salts, preferrably to be taken from the chiral pool, are (very) promising host resolving agents.

Table 9. Resolution of 2-Butanol (*132*) Using China Alkaloids *134–136* [46]

Host molecule	Configuration of the guest 2-butanol	ee [%]	Remarks
(−)-N-Methylquininiumiodide (*134*)	(+)-(*S*)	16–17	
(+)-N-Benzylquininiumchloride (*135*)	(−)-(*R*)	24–26	
		50	after twice recrystallization
(+)-N-Methylquinidiniumiodide (*136*)	(−)-(*R*)	16–17	

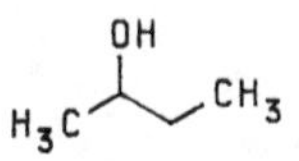

132

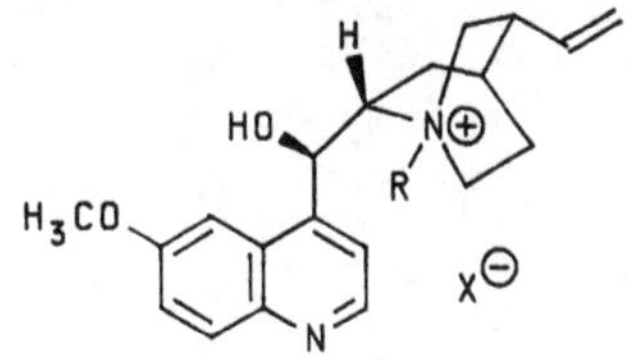

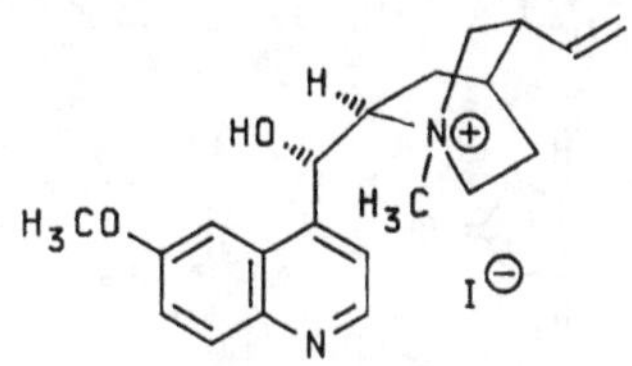

136

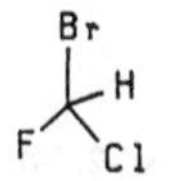

133

	R	X
134	CH_3	I
135	$CH_2C_6H_5$	Cl

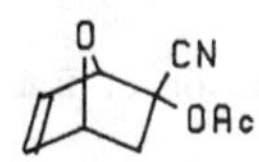

137

2.6 Molecular Architecture: Tailor-made Host Compounds

Beside the application of chiral natural products, today the synthesis of tailor-made host molecules to be used for racemate resolution is to the fore.

As the inclusion capacity of a host compound cannot be precisely predicted as yet, the work of the preparative chemist at the time being is reduced to systematic variation, design and further development of substance families, which are known to constitute clathrate formers. The numerous complex forming agents derived from binaphthol have been mentioned already in section 2.4.

Here, the chiral clathrate former 1,6-bis(*o*-chlorophenyl)-1,6-diphenylhexa-2,4-diyn-1,6-diol (*138*), derived from1,1,6,6-tetraphenylhexa-2,4-diyn-1,6-diol and developed by Toda shall be presented. Using this host compound, numerous guest substances, e.g. cyclic ketones and lactones which constitute important synthetic building blocks, have been successfully separated into enantiomers [38, 55–57].

A further example of a tailor-made host compound is "*cryptophane*" (*162*) [41]. By use of this cyclophane derived from the triveratrylene skeleton and having a large cavity inside the molecule, the differentiation of bromochlorofluoromethane (*136*) was successfully carried out on an analytical scale. Bromochlorofluoromethane has also been optically enriched through complexation with brucine [40].

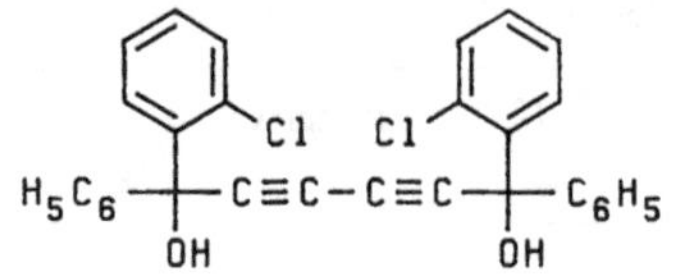

138

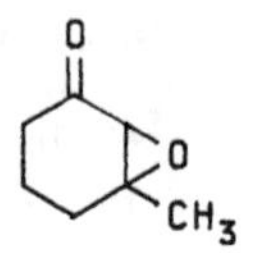

139

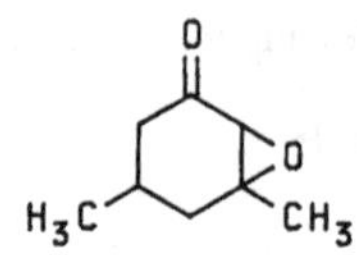

140

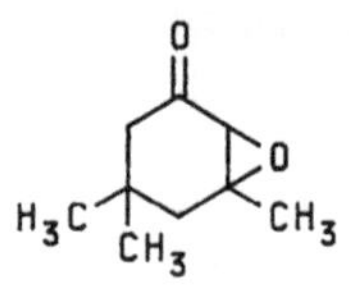

141

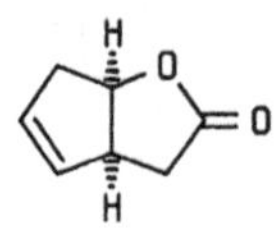

142

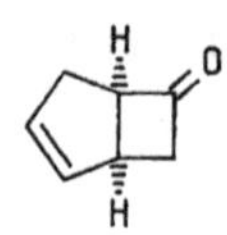

143

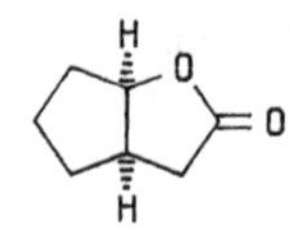

144

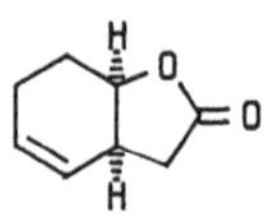

145

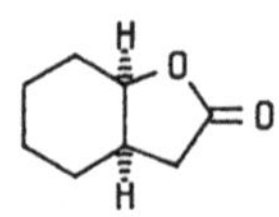

146

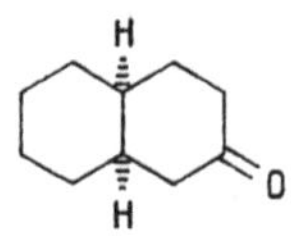

147

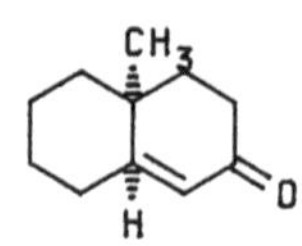

148

Table 10. Resolution of Racemates Using (—)-1,6-Bis(*o*-chlorophenyl)-1,6-diphenylhexa-2,4-diyn-1,6-diol (*138*)

Guest molecule	ee [%]	$[\alpha]_D$	Configuration (opt. rotation)	Remarks
3-Methyl-2,3-epoxycyclohexanone (*139*)	100	58.3	(+)	after recrystallization (3 times)
3,4-Dimethyl-2,3-epoxycyclohexanone (*140*)	90	−122	(—)	
	100	−136	(—)	after recrystallization (3 times)
3,5,5-Trimethyl-2,3-epoxycyclohexanone (*141*)	100	13.5	(+)	after recrystallization (3 times)
6-Oxabicyclo[3.3.0]oct-2-en-7-one (*142*)	8.5		(+)	
	100	110	(+)	after recrystallization (10 times)
Bicyclo[3.2.0]hept-2-en-6-one (*143*)	8.3	−5.2	(—)	
2-Oxabicyclo[3.3.0]octan-3-one (*144*)	20		(+)	
	100	13.3	(+)	after recrystallization (5 times)
7-Oxabicyclo[4.3.0]non-2-en-8-one (*145*)	100		(+)	after recrystallization (3 times)
7-Oxabicyclo[4.3.0]nonan-8-one (*146*)	100	50.3	(+)	after recrystallization (3 times)
Bicyclo[4.4.0]decan-3-one (*147*)	100	−3.9	(—)	after recrystallization (3 times)
6-Methylbicyclo[4.4.0]dec-1-en-3-on (*148*)	100	218	(+)	after recrystallization (3 times)

Table 10. (continued)

Guest molecule	ee [%]	$[\alpha]_D$	Configuration (opt. rotation)	Remarks
3-Methylcyclohexanone (*149*)	28	4.0	(+)	
	66	9.5	(+)	after recrystallization (3 times)
	100	14.4	(+)	after recrystallization (5 times)
3-Methylcyclopentanone (*150*)	100	−148	(−)	after recrystallization (7 times)
5-Methyl-γ-butyrolactone (*151*)	100	30.1	(+)	after recrystallization (12 times)
1-Hydroxycyclopent-2-en-4-one (*152*)	100	— no remarks —		via *153–155*
n-Butanoic acid cyclopent-2-en-4-on-1-ol ester (*153*)	100	−134	(−)	
n-Pentanoic acid cyclopent-2-en-4-on-1-ol ester (*154*)	100	— no remarks —		
Dimethylpropanoic acid cyclopent-2-en-4-on-1-ol ester (*155*)	100	— no remarks —		
n-Heptylmethylsulfoxide (*156*)	100	105	(+)	
n-Hexylmethylsulfoxide (*157*)	100	118	(+)	
n-Pentylmethylsulfoxide (*158*)	100	99.5	(+)	
n-Butanoic acid (2,3-epoxy-1-propyl)ester (*159*)		— no remarks —		
Acetic acid tetrahydrofurylmethyl-ester (*160*)		— no remarks —		
Acetic acid tetrahydropyrylmethyl-ester (*161*)		— no remarks —		

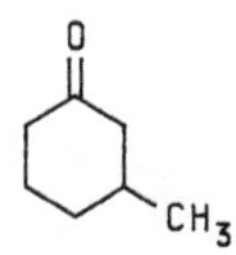

149

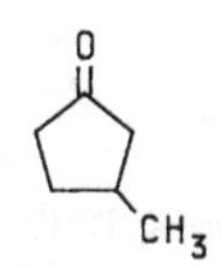

150

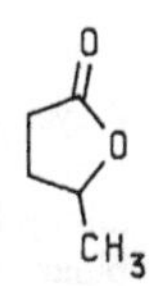

151

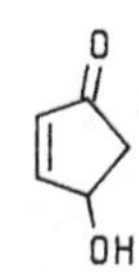

152

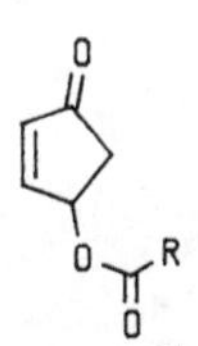

	R
153	n-C_3H_7
154	n-C_4H_9
155	*tert*-C_4H_9

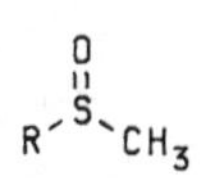

	R
156	C_7H_{15}
157	C_6H_{13}
158	C_5H_{11}

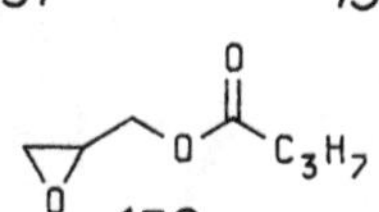

159

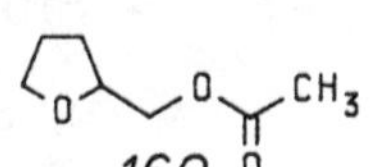

160

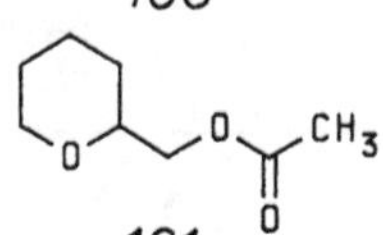

161

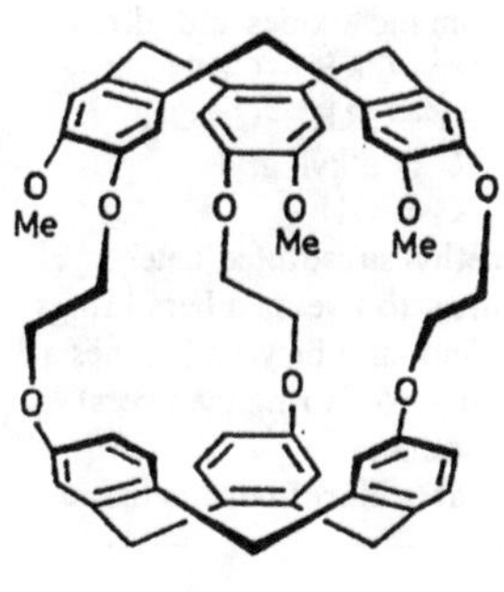

162

3 Conclusion and Outlook

The examples of racemate resolution by clathrate formation described in this review, demonstrate in how many ways host compounds can be applied today in this branch of chemistry. The experimental chemist, who faces the task of finding the matching host substance for the enantiomeric enrichment of a specific guest compound, might think that success in this field is determined by chance. Indeed, many (if not most) clathrates have been found by chance or in the course of a large series of experiments. With the increasing number of clathrates studied, nevertheless, criteria can be worked out which facilitate the search for an appropriate host compound, being restricted to a relatively small number of substances.

Spatial requirements use to play a decisive role in host/guest interactions as well as number and position of polar centres of the guest molecule. In Table 11 the attempt has been made to list some guest molecules according to these aspects. For the chemist this might facilitate the choice of appropriate host compounds for given guest substances. In future, new, more specific host compounds will be produced more quickly than hitherto and the chemist will find an appropriate host substance for the enantiomer differentiation and resolution of a given guest substance more easily.

Table 11. Host Compounds Suitable for the Resolution of Racemates of Some Types of Guest Substances

Guest	Host
long chained esters R—COOR′	urea
(R, R′ = alkyl C_5–C_{10})	
substituted n-alkanes	"TOT"
$H_3C–(CH_2)_n–CH(X)–(CH_2)_m–CH_3$	quaternary quininium salts (n = 0, m = 1)
X = Cl, Br, OH, NH_2	
n = 0, 1	
m = 1–9	
aryl- and cyclohexyl acetic acid esters	β-cyclodextrin
Ar—CRR′—COOR″	
sulfoxides,	α,β-cyclodextrins, TOT,
sulfinic acid esters,	binaphthyl, 1,6-bis(*o*-chlorophenyl)-
phosphinic acid esters	1,6-diphenylhexa-2,4-diyn-1,6-diol
aromatic alkines and nitrils	brucine, sparteine
Ar—CRR′—C≡N	
Ar—CRR′—C≡CH	
R = alkyl, aryl	
R′ = OH	
methyl substituted heterocyclic	"TOT"
three- to five-membered rings	
mono- and bicyclic ketones and lactones	1,6-bis(*o*-chlorophenyl)-1,6-diphenyl-
(three to six ring members)	hexa-2,4-diyn-1,6-diol
halothane	β-cyclodextrin
bromochlorofluoromethane	brucine, "cryptophane" [41]

4 References

1. Weber, E., Josel, H.-P.: J. Incl. Phenom. *1*, 79 (1983)
2. Schlenk, Jr., W.: Liebigs Ann. Chem. *1973*, 1145
3. Schlenk, Jr., W.: Liebigs Ann. Chem. *1973*, 1179
4. Newman, A. C. D., Powell, H. M.: J. Chem. Soc. *1952*, 3747
5. Edge, S. J., Ollis, W. D., Stephanatou, J. S. et al.: Tetrahedron Lett. *22*, 2229 (1981)
6. Arad-Yellin, R., Green, B. S., Knossow, M. et al.: J. Am. Chem. Soc. *105*, 4561 (1983)
7. Arad-Yellin, R., Green, B. S., Knossow, M.: J. Am. Chem. Soc. *103*, 1157 (1980)
8. Gerdil, R., Allemand, J.: Helv. Chim. Acta *63*, 1750 (1980)
9. Allcock, H. R.: Acc. Chem. Res. *11*, 81 (1978)
10. Bishop, R., Dance, I., Hawkins, S. C.: J. Chem. Soc., Chem. Commun. *1983*, 889
11. MacNicol, D. D., Mallinson, P. R., Murphy, A. et al.: Tetrahedron Lett. *23*, 4131 (1982)
12. Hardy, H. D. U., MacNicol, D. D., Wilson, D. R.: J. Chem. Soc., Perkin Trans. 2, *1979*, 1011
13. Freer, A., Gilmore, G. J., MacNicol, D. D. et al.: Tetrahedron Lett. *21*, 205 (1980)
14. MacNicol, D. D., Mc Kendrick, J. J., Wilson, D. R.: Quart. Chem. Soc. Rev. *7*, 65 (1978)
15. Tsadon, B., Decsei, L., Szilasi, M. et al.: J. Chromatogr. *270*, 127 (1983)
16. Koscielski, T., Sybilska, D., Jurczak, J.: J. Chromatogr. *280*, 131 (1983)
17. Knabe, J., Agarwal, N. S.: Dtsch. Apoth.-Ztg. *113* (38), 1449 (1973)
18. Benschop, H. P., van den Berg, G. R.: J. Chem. Soc., Chem. Commun. *1970*, 1431
19. Cramer, F., Dietsche, W.: Chem. Ber. *92*, 378 (1959)
20. Mikolajczyk, M., Drabowicz, J.: Tetrahedron Lett. *1972*, 2379
21. Yamanari, K., Shimura, Y.: Chem. Lett. *1982*, 1957
22. Mikolajczyk, M., Drabowicz, J.: J. Am. Chem. Soc. *100*, 2510 (1978)
23. Dotsevi, G., Sogah, G. D. Y., Cram, D. J.: J. Am. Chem. Soc. *98*, 3038 (1976)
24. Cram, D. J., Helgeson, R. C., Peacock, S. C. et al.: J. Org. Chem. *43*, 1930 (1978)
25. Gokel, G. W., Timko, J. M., Cram, D. J.: J. Chem. Soc., Chem. Commun. *1975*, 394
26. Cram, D. J., Helgeson, R. C., Sousa, L. R. et al.: Pure Appl. Chem. *43*, 327 (1975)
27. Lingenfelter, D. S., Helgeson, R. C., Cram, D. J.: J. Org. Chem. *46*, 393 (1981)
28. Peacock, S. C., Walba, D. M., Gaeta, F. C. A. et al.: J. Am. Chem. Soc. *102*, 2043 (1980)
29. Sousa, L. R., Dotsevi, G., Sogah, G. D. Y. et al.: J. Am. Chem. Soc. *100*, 4569 (1978)
30. Peacock, S. C., Domeier, L. A., Gaeta, F. C. A. et al.: J. Am. Chem. Soc. *100*, 8190 (1978)
31. Timko, J. M., Helgeson, R. C., Cram, D. J.: J. Am. Chem. Soc. *100*, 2828 (1978)
32. Dotsevi, G., Sogah, G. D. Y., Cram, D. J.: J. Am. Chem. Soc. *101*, 3035 (1979)
33. Newcomb, M., Toner, J. L., Helgeson, R. C., Cram, D. J.: J. Am. Chem. Soc. *101*, 4941 (1979)
34. Sousa, L. R., Hoffman, D. H., Kaplan, L., Cram, D. J.: J. Am. Chem. Soc. *96*, 7100 (1974)
35. Kyba, E. P., Gokel, G. W., de Jong, F. et al.: J. Org. Chem. *42*, 4173 (1977)
36. Peacock, S. C., Cram, D. J.: J. Chem. Soc., Chem. Commun. *1976*, 282
37. Curtis, W. D., King, R. M., Stoddart, J. F. et al.: J. Chem. Soc., Chem. Commun. *1976*, 284
38. Toda, F.: this volume, p. 43
39. Toda, F., Tanaka, K., Nagamatsu, S.: Tetrahedron Lett. *25*, 4929 (1984)
40. Wilen, S. H., Bunding, K. A., Kascheres, C. M. et al.: J. Am. Chem. Soc. *107*, 6997 (1985)
41. Canceill, J., Lacombe, L., Collet, A.: J. Am. Chem. Soc. *107*, 6993 (1985)
42. Toda, F., Tanaka, K.: Tetrahedron Lett. *22*, 4669 (1981)
43. Toda, F., Tanaka, K., Ueda, H. et al.: J. Chem. Soc., Chem. Commun. *1983*, 743
44. Toda, F., Tanaka, K., Mori, K.: Chem. Lett. *1983*, 827
45. Toda, F., Tanaka, K.: Chem. Lett. *1983*, 661
46. Worsch, D., Vögtle, F., Kirfel, A., Will, G.: Naturwissenschaften *71*, 423 (1984)
47. Löhr, H.-G., Vögtle, F., Puff, H., Schuh, W.: J. Chem. Soc., Chem. Commun. *1983*, 924
48. Vögtle, F., Löhr, H.-G., Puff, H., Schuh, W.: Angew. Chem. *95*, 425 (1983); Angew. Chem., Int. Ed. Engl. *22*, 409 (1983); Angew. Chem. Suppl. *1983*, 527–536
49. Löhr, H.-G., Josel, H.-P., Engel, A., Vögtle, F. et al.: Chem. Ber. *117*, 1487 (1984)
50. Löhr, H.-G., Vögtle, F., Schuh, W., Puff, H.: J. Incl. Phenom. *1*, 175 (1983)
51. Worsch, D., Vögtle, F.: J. Incl. Phenom. *4*, 163 (1986)
52. Vögtle, F., Löhr, H.-G., Franke, J., Worsch, D.: Angew. Chem. *97*, 721 (1985); Angew. Chem., Int. Ed. Engl. *24*, 727 (1985)
53. Weber, E., Müller, U., Worsch, D., Vögtle, F. et al.: J. Chem. Soc., Chem. Commun. *1985*, 1578

Isolation and Optical Resolution of Materials Utilizing Inclusion Crystallization

Fumio Toda

Department of Industrial Chemistry, Faculty of Engineering, Ehime University, Matsuyama 790, Japan

Table of Contents

When a host molecule recognizes the shape of a guest molecule efficiently and includes one isomer of a mixture selectively, the process is usable for compound separation. Furthermore, when an optically active host molecule recognizes the chirality of a guest molecule efficiently and includes one optical isomer selectively, its behaviour is good for optical resolution.

In these studies, an important subject is related to the design of efficient host compounds. Using our simple idea, we designed a series of host molecules available for the isolation and optical resolution of different guest species. Some examples of isolation and optical resolution making use of these hosts are described.

1 Introduction

This chapter describes the isolation and optical resolution of materials utilizing crystalline inclusion complex formation covering studies which have been carried out by our research group since 1985.

2 Design of Host Compounds

Until now, we have synthesized more than thirty new host compounds by employing a novel but simple idea, and the number of them is increasing. This section deals with the design of such compounds, which are tentatively classified into three groups.

If newly designed host compounds are chiral and can be resolved, they are useful for optical resolution of guest species. The preparation of such hosts is described in section 4.

2.1 Diol Host Compounds

In 1968, we found that 1,1,6,6-tetraphenylhexa-2,4-diyne-1,6-diol (*1a*) and 1,1,4,4-tetraphenylbut-2-yne-1,4-diol (*1b*) form crystal inclusions with a wide variety of guest compounds, as shown in Table 1 [1,2]. In most cases, *1a* and *1b* include guest species in 1:2 and 1:1 ratios, respectively. The νOH absorption band of all inclusion compounds appears at lower frequencies than those of uncomplexed *1a* and *1b* [1,2]. This suggests the presence of strong hydrogen bonding between host and guest (inclusion "complexes").

Table 1. Some typical guest compounds which are included in *1a–c* and *19a–c*, and their molar ratios to the host compounds

Guest compound	Host compound					
	1a	*1b*	*1c*	*19a*	*19b*	*19c*
Methanol	1	—	—	—	—	—
Ethanol	1	—	—	—	—	—
Acetone	2	1	2	—	—	—
Cyclopentanone	2	1	2	—	2	1
γ-Butyrolactone	2	1	2	—	—	—
Benzaldehyde	2	—	—	—	—	—
Tetrahydrofuran	2	1	1	—	—	—
Dioxane	2	1	1	1	1	1
Carbon tetrachloride	0.5	—	1	—	—	—
Dimethylformamide	2	2	2	—	2	1
Dimethyl sulfoxide	2	2	2	0.5	2	0.5
Acetonitrile	2	1	—	—	2	2
Benzene	0.5	—	1	2	2	1
Pyridine	2	1	2	—	2	1

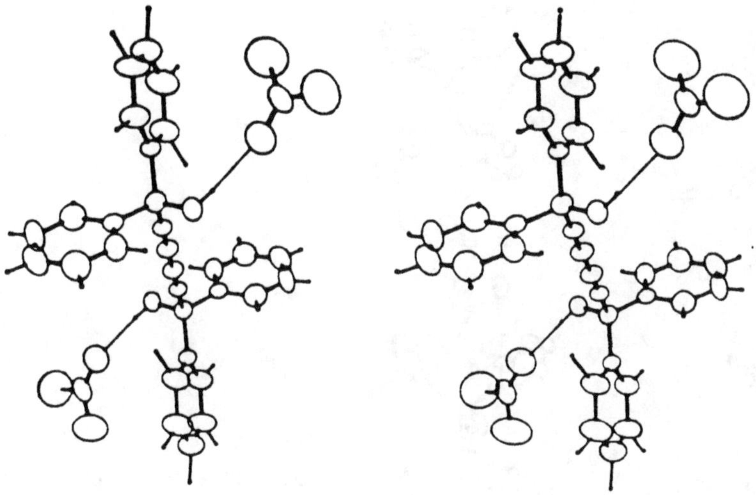

Fig. 1. Stereodrawing of the 2:1 acetone crystal inclusion of *1a* (thin lines represent H-bond contacts, O atoms dotted)

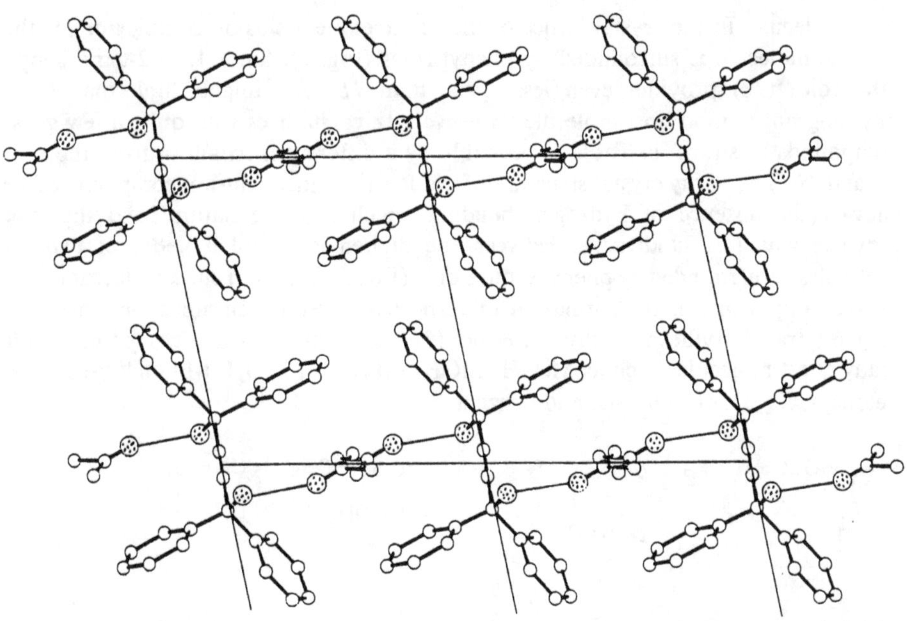

Fig. 2. Packing diagram of the 1:1 acetone crystal inclusion of *1b*. (thin lines represent H-bond contacts, O atoms dotted)

The X-ray crystal structural study of the acetone inclusion complexes of *1a* [3] and *1b* [2] disclosed that hydrogen bonds between the hydroxyl groups of *1a* and *1b*, respectively, and the carbonyl oxygen of acetone play an important role (Figs. 1 and 2). Although *1a* has enough space between the two diphenylcarbinol moieties to include two acetone molecules, *1b* has only sufficient space to accomodate one ace-

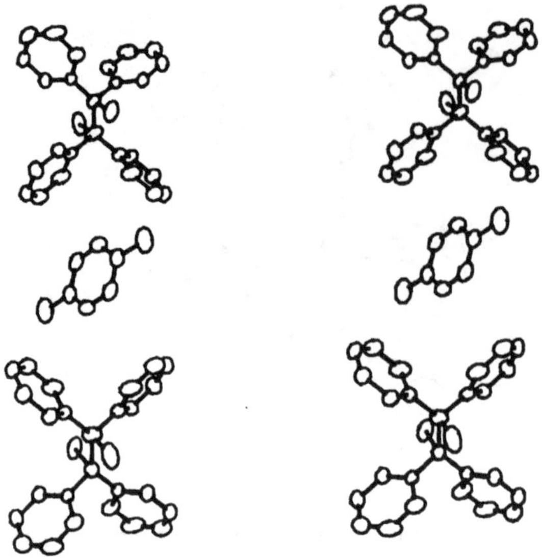

Fig. 3. Stereodrawing of the 1:2 p-xylene inclusion compound of *1c* (O atoms dotted)

tone molecule. In the crystal lattice of the 1:1 acetone inclusion complex of *1b*, the acetone molecule is surrounded by phenyl rings (Fig. 2). Since 1,1,2,2-tetraphenylethanediol (*1c*) provides even less space than *1b*, one might think that *1c* is anything but a good host molecule. However, *1c* crystallizes with quite a few guest compounds as shown in Table 1, probably by a different mechanism from those of *1a* and *1b*. The X-ray crystal structure of the 1:1 p-xylene inclusion compound of *1c* showed the absence of hydrogen bonding which is quite natural and that the p-xylene which is sandwiched between the diphenylcarbinol moieties of two *1c* molecules is surrounded by phenyl groups only (Fig. 3). This surrounding factor might be very important in the formation of respective crystal inclusions. For instance, 1,1,6,6-tetra-t-butylhexa-2,4-diyne-1,6-diol (*2*) also forms inclusion compounds with many guest molecules such as MeOH, EtOH, n-PrOH, i-PrOH, t-BuOH, γ-butyrolactone, CCl_4, DMF, DMSO, and benzene.

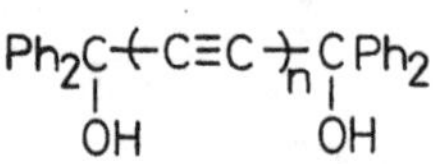

a: n = 2
b: n = 1
c: n = 0

1

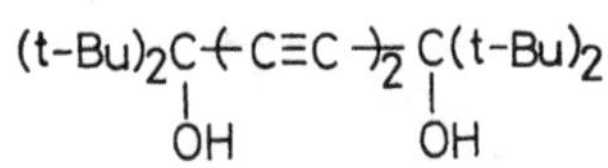

2

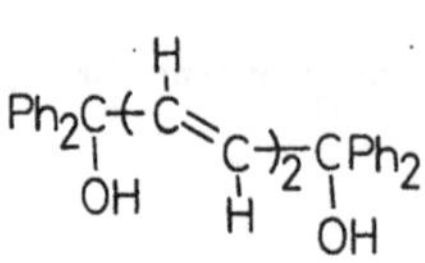

3

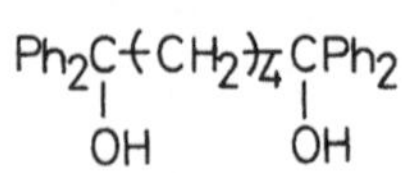

4

The rigidity of the molecular constitutions of *1* and *2* seems also an important fact, since (*E*, *E*)-1,1,6,6-tetraphenylhexa-2,4-diene-1,6-diol (*3*) includes only a few guest compounds in the crystal and 1,1,6,6-tetraphenylhexane (*4*) does not form any inclu-

Table 2. Some typical guest compounds which are included in *5*, *6a*, *7*, *8a*, *8b*, *9a*, *9b*, *10a*, *10b*, *11*, and *20*. Numbers specify molar ratios to the host compounds

Guest compound	Host compound										
	5	*6a*	*7*	*8a*	*8b*	*9a*	*9b*	*10a*	*10b*	*11*	*20*
Methanol	1	2	—	—	—	—	—	—	—	1	—
Ethanol	2	—	—	2	—	—	—	—	—	1	—
Acetone	2	2	—	2	—	1	—	—	—	1	—
Cyclopentanone	2	2	2	—	2	1	—	2	1	1	—
γ-Butyrolactone	2	—	2	—	2	—	—	—	1	1	1
Benzaldehyde	2	2	—	—	—	—	—	—	—	—	—
Tetrahydrofuran	2	—	2	2	1	2	1	—	1	1	—
Dioxane	2	2	1	1	—	—	—	—	1	1	1
Carbon tetrachloride	—	—	—	—	2	1	—	—	1	1	—
Dimethylformamide	2	2	2	2	2	2	—	—	—	1	—
Dimethyl sulfoxide	2	2	2	2	—	2	—	—	—	1	—
Acetonitrile	2	2	—	—	—	2	—	—	—	1	—
Benzene	—	—	—	—	1	—	1	2	1	1	1
Pyridine	2	2	2	—	1	2	—	—	1	1	1

sion compound at all. These observations showed us a way for the design of new potential host compounds, namely, using a rigid molecule which has an *anti*-diol function and some bulky hydrophobic groups such as phenyl. According to this idea, we first designed 9,10-dihydroxy-9,10-diphenyl-9,10-dihydroanthracene (*5*), 2,5-diarylhydroquinones (*6*), and 2,2′-dihydroxy-1,1′-binaphthyl (*7*), and found them to be good inclusion hosts [5]. The molar ratios of the inclusion compounds of *5*–*7* with

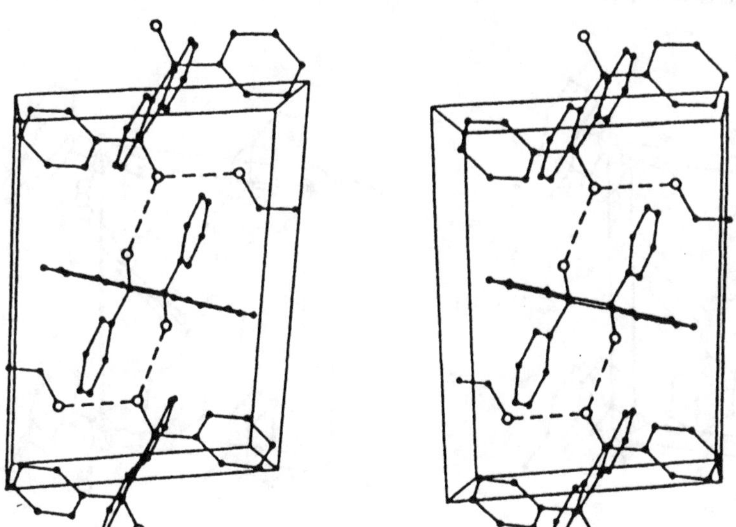

Fig. 4. Stereodrawing of the 1:2 ethanol inclusion compound of *5* (broken lines represent H-bond contacts; O atoms as circles, C atoms as bold dots)

typical guest molecules are shown in Table 2 (For additional inclusion properties of *7* see volume 2 of this topic, Chapter Weber).

By the X-ray crystal structural studies of the inclusion compounds of *5* with ethanol (Fig. 4)[6-8], *6b* with ethanol (Fig. 5)[7], and *7* with methyl m-tolyl sulfoxide (Fig. 6)[9], the presence of a twisted relationship between two aryl groups was found in all cases. For example, phenyl and benzo groups in *5*, 2,4-dimethylphenyl and dihydroxybenzene groups in *6b*, and two naphthyl groups in *7* are twisted with

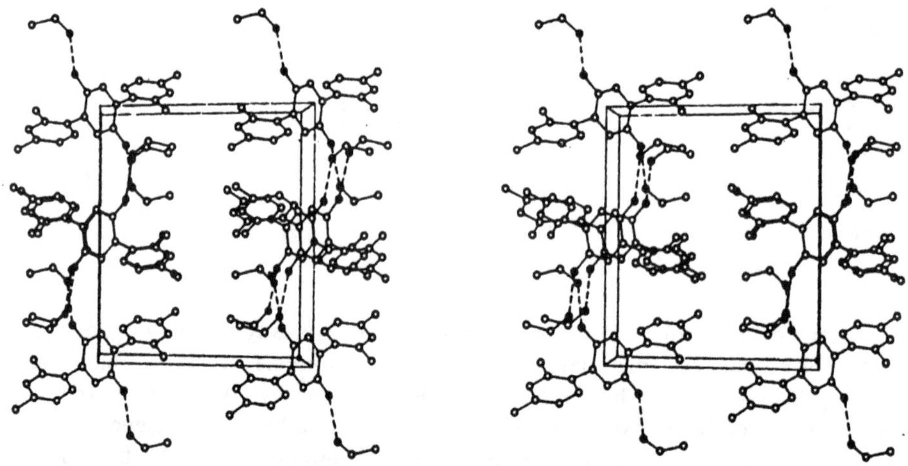

Fig. 5. Stereodrawing of the 1:2 ethanol of *6b* (broken lines represent H-bond contacts; C atoms as circles, O atoms as bold dots)

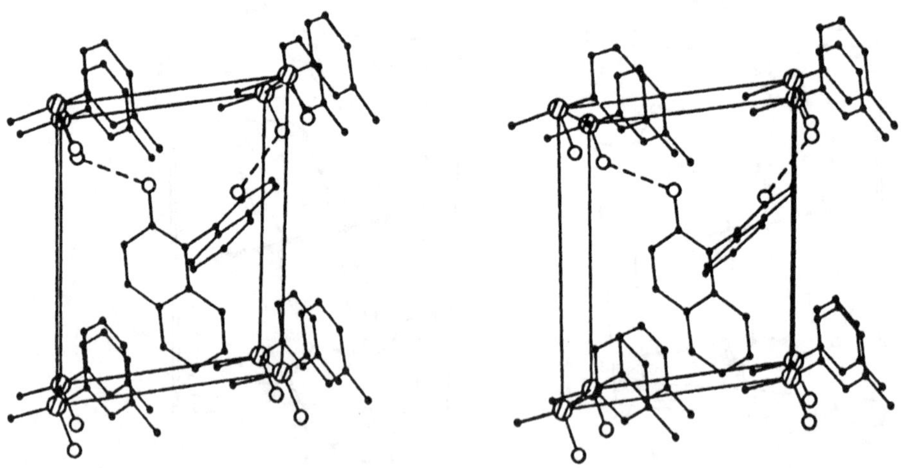

Fig. 6. Stereodrawing of the 1:2 methyl (R)-(+)-m-tolyl sulfoxide crystal inclusion of (R)-(+)- (broken lines represent H-bond contacts; O atoms as circles, C atoms as bold dots, S atoms hatched)

respect to each other. In the latter two cases, two aromatic groups are almost perpendicular to each other. Compounds *5–7* have also two hydroxyl groups in an *anti*-position which is supposed to be a prerequisite for hydrogen bonded inclusion complexes.

We then poceeded in designing host compounds with aromatic moieties in a twisted position and provided with two hydroxyl groups. Examples are 2,2′-dihydroxy-9,9′-dianthryl- (*8a*), 9,9′-dihydroxy-10,10′-biphenanthryl (*9a*), and 2,2′-dihydroxy-9,9′-spirobifluorene (*10a*)[10]. The inclusion compounds obtained of these hosts with some typical guests are given in Table 2 (cf. also volume 2 of this topic, Chapter Weber). Following the same strategy, we came to another host compound, 1,1-di(p-hydroxyphenyl)-cyclohexane (*11*) (Table 2)[10]. The twisted host molecules, *7*, *8a*, *9a*, and *10a* are chiral. Hence their optically active isomers can be used for optical resolution of guest compounds (see Sect. 4.3).

A further finding is that the prop-2-yn-1-ol derivatives *12–14* form crystal inclusions with alkaloids such as brucine and sparteine which readily allowed optical resolution of *12–14*[11–13]. By the same method, the cyanohydrin *15*[14] and the simple secondary alcohols *16*[10] were also easily resolved. The optical resolution procedures are described in Section 4.1.

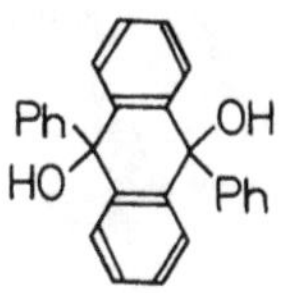

5

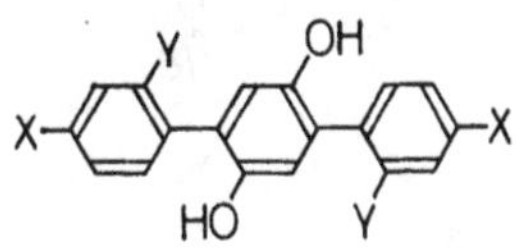

6 *a*: X=Y=H
b: X=Y=Me
c: X=Cl; Y=H

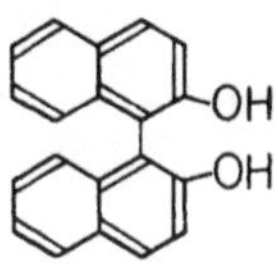

7

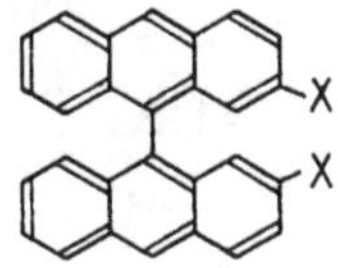

8

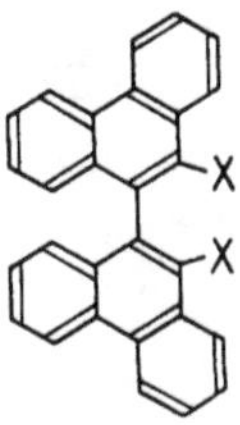

9

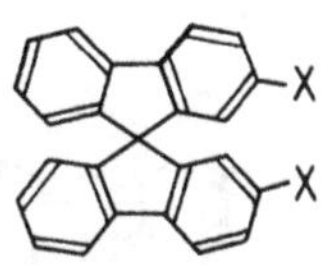

10

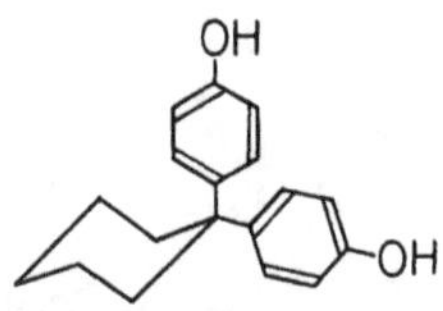

11

a: X = OH
b: X = H

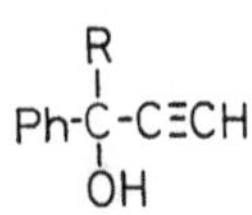

12

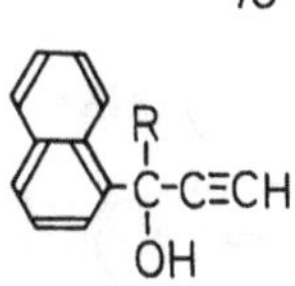

13

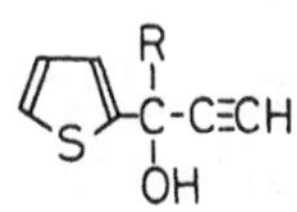

14

Of the prop-2-yn-1-ol derivatives, 1,1-bis(2,4-dimethylphenyl)prop-2-yn-1-ol *17* and 2-methylbut-3-yn-2-ol *18* were found to form stable inclusion complexes with ethanol and alkali metal hydroxides, respectively. These phenomena can be used for the separation of ethanol and alkali metal hydroxides from their aqueous solutions (for a detailed discussion see Sect. 3).

2.2 Hydrocarbon Host Compounds

As described before, the p-xylene inclusion compound of *1c* shows the guest molecule fixed by the surrounding phenyl groups of *1c* only, and hydrogen bonding plays no role[4]. This result prompted us to design new hosts without hydroxyl groups.

At first, the hydroxyl groups of *1a–c* were replaced by hydrogen giving the corresponding hydrocarbon hosts, 1,1,6,6-tetraphenylhexa-2,4-diyne (*19a*), 1,1,4,4-tetraphenylbut-2-yne (*19b*), and 1,1,2,2-tetraphenylethane (*19c*). They form crystal inclusions with guests as given in Table 1. The same idea was applied to *8a*, *9a*, and *10a* leading to 9,9′-bianthryl (*8b*), 9,9′-biphenanthryl (*9b*), and 9,9′-spirobifluorene (*10b*). They also allow inclusion formation with some guest compounds (Table 2). Furthermore, tryptycene *20* was also found to be a good host candidate (Table 2). However, 1,1′-binaphthyl does not include any guest compound under the chosen conditions (contrast with volume 2 of this topic, Chapter Weber).

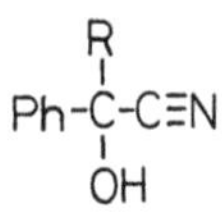

15

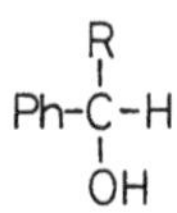

16

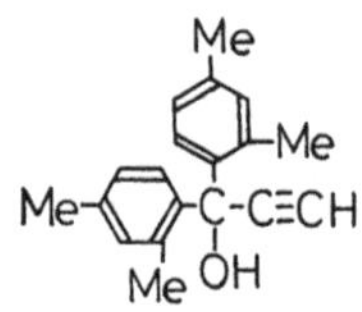

17

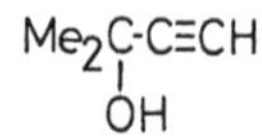

18

a : R = t-Bu

b : R = CCl_3

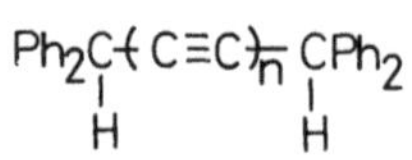

19

a: n = 2

b: n = 1

c: n = 0

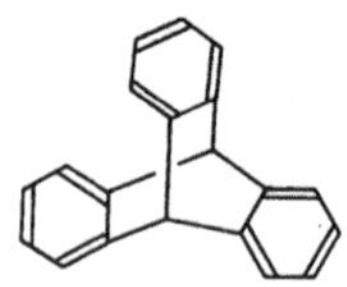

20

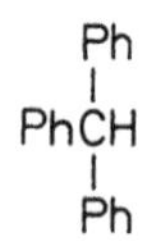

21

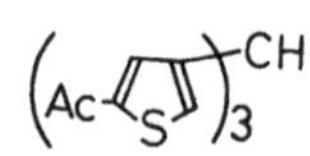

22

$(i\text{-}Pr)_2NCOCON(i\text{-}Pr)_2$

23

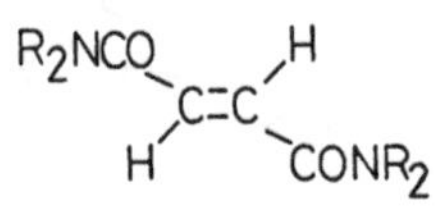

24

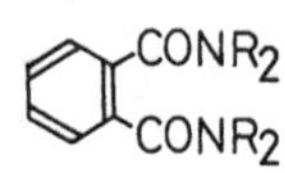

25

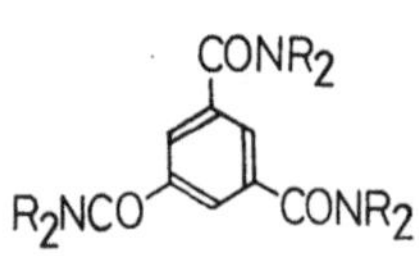

26

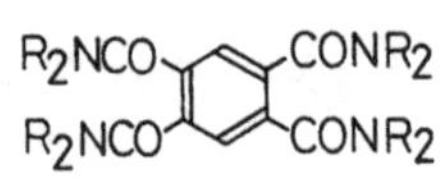

27

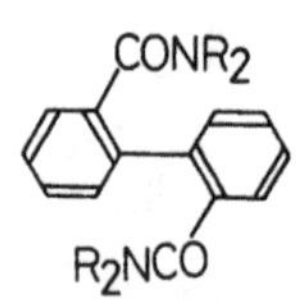

28

We then searched for simple hydrocarbon hosts which have a twisted structure of aromatic groups and found that triphenylmethane *21* forms 1:1 inclusion compounds with benzene, toluene, and dioxane. In contrast, tetraphenylmethane was found to be a poor host. Only carbon tetrachloride is included in an 1:1 ratio, probably because of over-crowing of the four phenyl groups. Recently, we noticed that *21* [14] as well as tris(5-acetyl-3-thienyl)methane *22* [15] have been recognized already as host compounds.

2.3 Amide Host Compounds

Amides form stable hydrogen-bonded inclusion complexes with *1a* as depicted in Fig. 7, I [1]. This in turn suggests that a diamide should possibly work as a good host molecule for alcohols (Fig. 7, II). According to this idea, we designed the diamides

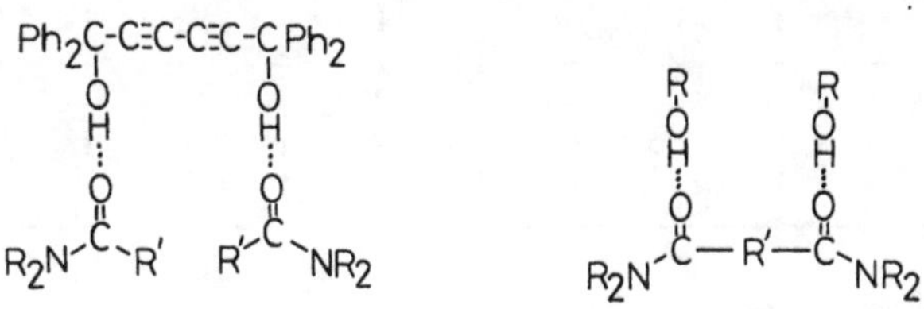

I II

Fig. 7. Suggested binding modes of carbonamide alcohol crystal inclusions (mutual recognition)

Table 3. Alcohols which are included in *23–32*, and their molar ratios to the host compounds

Guest compound	Host compound									
	23	*24*	*25*	*26*	*27*	*28*	*29*	*30*	*31*	*32*
Methanol	—	3	1	2	4	1	1	1	1	1
Ethanol	—	1	—	1	—	1	1	1	1	—
1-Propanol	—	1	1	1	3	1	1	—	1	1
2-Propanol	—	2	1	1	3	1	1	1	1	1
1-Butanol	—	—	—	—	1	1	1	—	—	1
2-Butanol	—	—	2	1	1	1	1	—	—	1
Isobutanol	—	1	1	1	1	1	1	—	1	1
t-Butanol	—	2	1	—	2	1	1	1	1	1
1-Decanol	—	—	—	—	1	—	—	—	—	—
Propargyl alcohol	—	1	1	1	3	1	1	1	1	—
Cyclohexanol	—	—	—	1	2	1	2	—	—	—
Ethylene glycol	—	0.5	0.5	1	—	1	1	1	0.5	1
1,3-Propanediol	—	—	0.5	1	—	1	1	—	—	—
1,4-Butanediol	—	0.5	—	—	—	0.5	2	1	0.5	1
1,5-Pentanediol	—	0.5	—	—	—	0.5	0.5	—	—	—
1,6-Hexanediol	—	0.5	—	—	—	0.5	0.5	—	—	—
1,8-Octanediol	—	0.5	—	—	—	—	0.5	—	—	—
Phenol	1	2	1	2	3	1	—	1	—	—
o-Cresol	2	2	2	2	1	1	—	2	—	—
m-Cresol	1	2	2	1	1	1	—	2	—	—
p-Cresol	2	2	2	2	3	2	—	2	1	—
Resorcinol	1	1	1	—	—	—	0.5	2	1	—

Table 4. Some typical guest compounds which are included in *23–32*, and their molar ratios to the host compounds

Guest compound	Host compound									
	23	*24*	*25*	*26*	*27*	*28*	*29*	*30*	*31*	*32*
Acetone	—	—	—	—	—	1	1	—	1	1
Cyclopentanone	—	—	1	0.5	3	1	2	1	—	—
γ-Butylrolactone	—	—	1	—	1	1	1	2	—	—
Benzaldehyde	—	—	—	1.5	1.5	0.5	—	—	1	—
Tetrahydrofuran	—	—	0.5	0.5	3	1	—	—	1	1
Dioxane	—	—	1	0.5	2	1	1	—	1	1
Carbon tetrachloride	—	—	—	1	3	1	—	—	1	1
Dimethylformamide	—	—	1	1	—	1	1	—	—	1
Dimethyl sulfoxide	—	—	—	—	—	1	1	—	—	—
Acetonitrile	—	—	—	—	—	—	2	—	—	—
Benzene	1	—	1	1.5	2	0.5	—	1	1	1
Pyridine	—	—	1	2	3	1	—	1	1	1

23–25, *28*, and *30–32*, the triamides *26* and *29*, and a tetraamide *27*, and found them to be good hosts for alcohols (Table 3) but also for ohter guest compounds (Table 4).

The results show that in all cases, R should be a larger group than i-Pr preferably a cyclohexyl group. Large R groups in the amide hosts would facilitate surrounding to the guest molecule in a crystalline lattice. Most amide hosts were found to be useful for the separation of isomers.

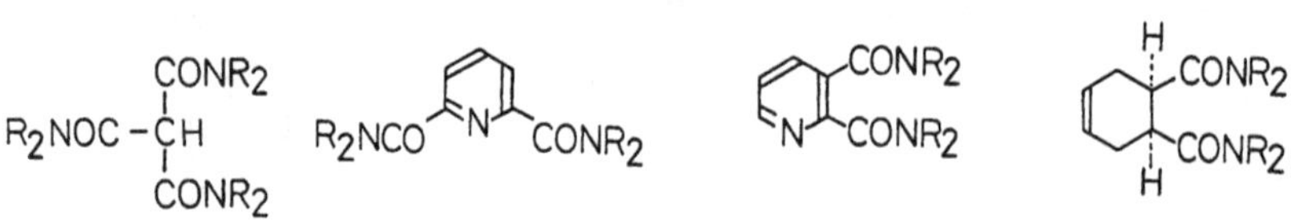

29 *30* *31* *32*

R = cyclohexyl

3 Isolation of Materials

Although the isolation and purification of materials are very important processes in laboratory and industry, these are not easy to carry out and sometimes almost impossible. If the method of inclusion compound formation can be used, the process then becomes very simple and cheap. Furthermore, this method is easily applicable to the separation of isomers having very close boiling points, since molecular recognition of host and guest is mostly due to a size and shape relationship.

3.1 Separation of Isomers

In some cases, isomers of aromatic, heteroaromatic, and aliphatic compounds are easily separated by inclusion formation with the host compounds described in Section 2.

For example, when a solution of *1a* and twice its molar amount each of o-(*33a*) and p-methylbenzaldehyde (*33b*) in ether light petroleum (1:2) was kept at room temperature for 12 h, a 1:2 inclusion compound of *1a* and p-isomer *33b* was formed as colorless crystals (mp 132 °C) in quantitative yield. Heating of the inclusion compound in vacuo gave 100% pure *33b* in 96% yield by distillation. The *1a* left after the distillation of *33b* can be used again. From the filtrate left after separation of the complex of *33b*, 99% pure o-isomer *33a* was obtained by distillation in vacuo in 90% yield [10]. This method can be applied to the separation of o- and p-isomers of *34–37* [10]. In most cases, only the p-isomer forms an inclusion compound with *1a*. In the case of *36* and *37*, however, both the o- and p-isomers form crystal inclusions

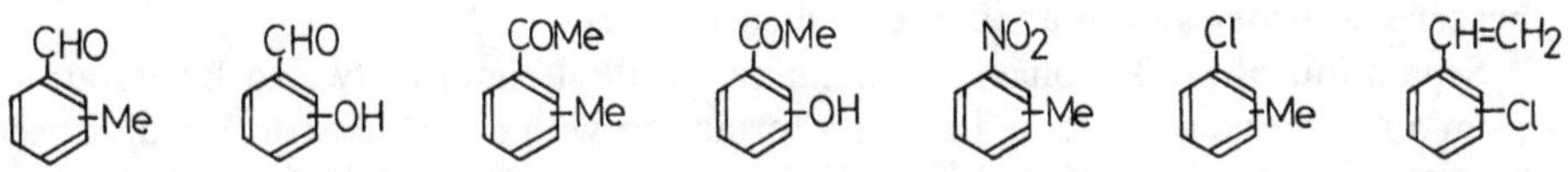

33 *34* *35* *36* *37* *38* *39*

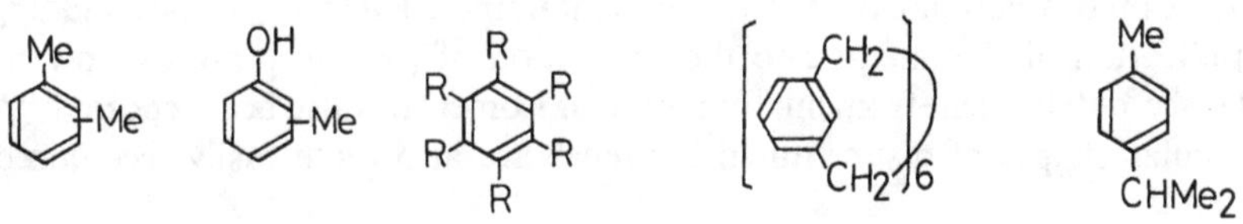

40 *41* *42* *43* *44*

a: ortho
b: para
c: meta

a: R=CH_2SCH_2Ph
b: R=CH_2OPh

a: ortho
b: para
c: meta

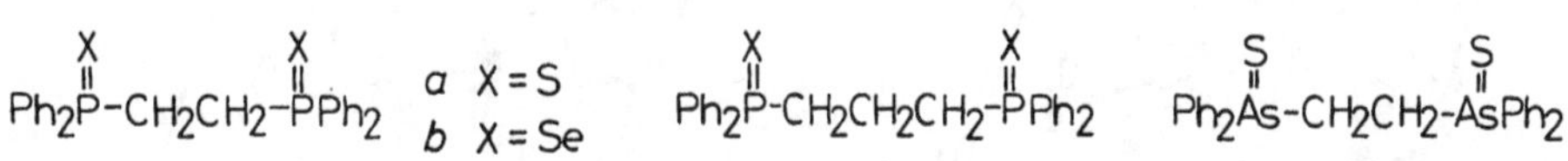

a X=S
b X=Se

45 *46* *47*

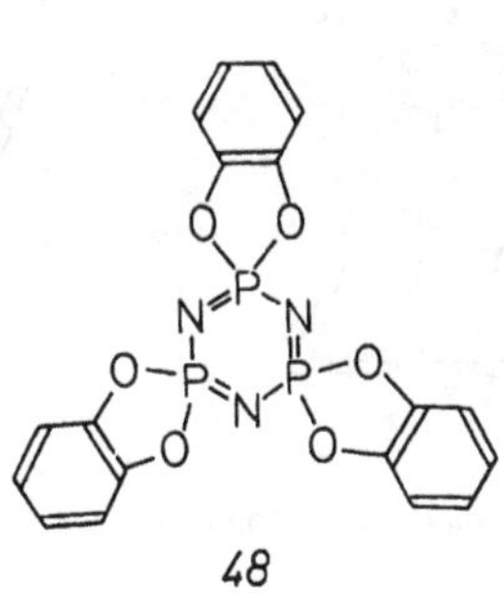

48

with *1a*. Nevertheless, they can be separated quite easily, because the solubilities of the inclusion compounds of o- and p-isomers in organic solvents are different.

Other diol hosts are also useful for the separation of guest isomers. As before, usually the p-isomers, e.g. of *38–40*, form inclusion compounds; thus they can be isolated from a mixture. In fact, o- (*40a*) and p-xylene (*40b*) were easily separated by *1c*. The X-ray crystal structure of the inclusion compound of *1c* with *40b* is shown in Fig. 3 [4].

Another important separation is that of m- (*41c*) and p-cresol (*41b*), for which the amide host *24* is effective [10]. p-Xylene (*40b*) was separated in 75% yield from its 1:1 mixture with o-isomer *40a* by crystal inclusion with *42*a. It is interesting that *42b* includes *40a* selectively [16].

Separation of *40b* and *41b* by inclusion formation with metacyclophane *43* has also been reported [17]. For example, heating of a solution of *43* (2 parts) and a 1:1:1:1 mixture of *40a*, *40b*, *40c*, and ethylbenzene (6 parts) in methanol (10 parts) at 100 °C resulted in an inclusion compound (2.2 parts), which upon heating gave *40b* (0.22 part). This method has been applied for separation of *37b*, *40b*, *41b*, p-toluidine, p-dichlorobenzene, p-chlorophenol, and several other p-disubstituted benzene derivatives from a mixture of their isomers.

Separation of p-alkyltoluenes containing an alkyl chain of two to five C atoms from an isomeric mixture by inclusion formation with α-cyclodextrin has also been reported [18]. For example, when a 35:27:38 mixture of o- (*44a*), m- (*44c*), and p-cymene (*44b*) was treated with an aqueous solution of 0.8 g α-cyclodextrin at room temperature for 1 h, a *44b* enriched α-cyclodextrin inclusion compound was obtained, which upon steam distillation gave *44b* (0.1 g) contaminated with 1% of *44a* and 2% of *44c*. Application of this method to the separation of p-ethyl, p-n-butyl, p-tert-butyl-, and p-sec-butyltoluene from mixtures of their isomers has also been reported [18].

The molecular shapes of disubstituted benzenes are also quite easily recognized

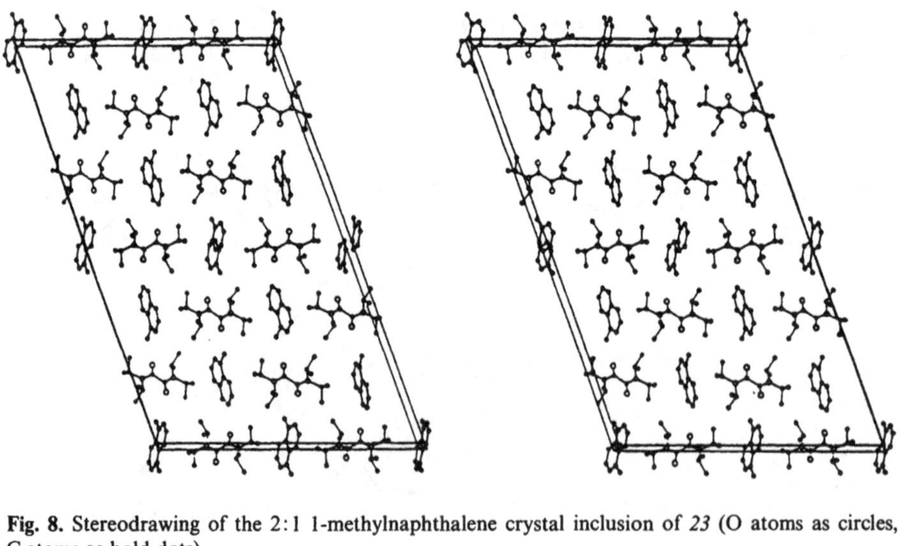

Fig. 8. Stereodrawing of the 2:1 1-methylnaphthalene crystal inclusion of *23* (O atoms as circles, C atoms as bold dots)

by some other unique hosts. 1,2-Bis(diphenylphosphinoselenoyl)ethane (*45a*) and its analogues *45b*, *46a–b*, and *47* were found to include various benzene derivatives [19]. The most profitable host of this sort (*45a*) exhibits a remarkably high selectivity for certain para-disubstituted benzenes. For example, recrystallization of *45a* from an equimolar mixture of each of the isomeric xylenes (*40a–c*) and ethylbenzene leads to incorporation of 97.5% of p-xylene (*40b*). The X-ray crystal structure of the inclusion compound between *45a* and *40b* (3:1) has been reported [19].

Similar selective inclusion phenomena related to the spiro host compounds tris(o-phenylenedioxy)cyclotriphosphazene (*48*) [20–22] and its 2,3-naphthalenedioxy- [20,21], 1,8-naphthalenedioxy [20,22] and 2,2'-diphenylenedioxy-analogues [20] have been studied by employing mass spectrometry, ^{1}H-NMR broadline, and crystal X-ray diffraction techniques [20–22].

Since 2-acetylnaphthalene (*50*) forms an inclusion compound with *1a* and the 1-isomer *49* does not, both isomers can be separated easily. 1-Methylnaphthalene (*51*, bp 240–243 °C) which forms an inclusion compound with *23* can be separated from the 2-isomer (*52*, bp 245 °C) which does not [23]. The X-ray crystal structure of the 1:1 inclusion of *51* with *23* (Fig. 8) showed that this compound is formed by van der Waals

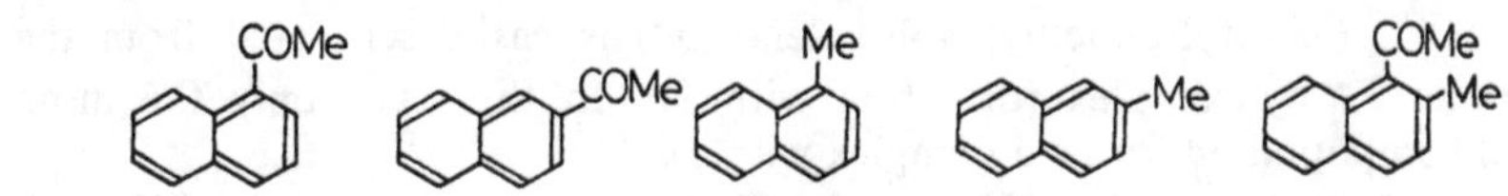

49 50 51 52 53

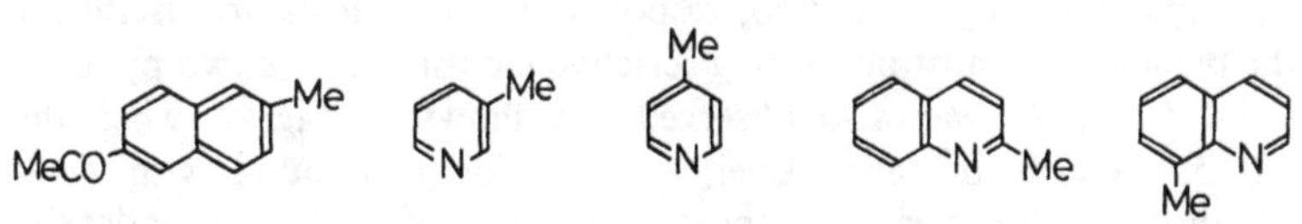

54 55 56 57 58

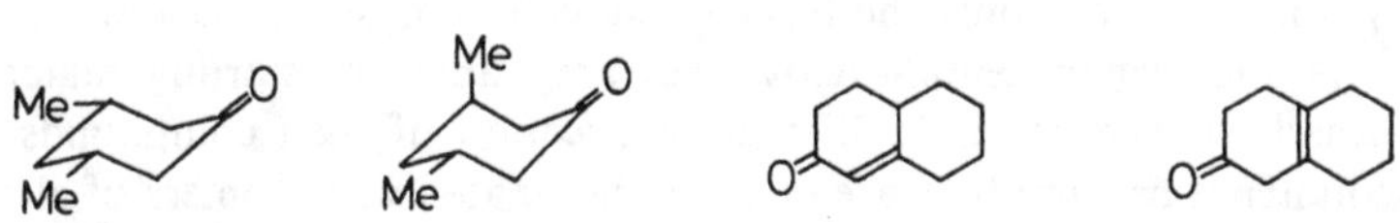

59 60 61 62

$HO{-}(CH_2)_4{-}OH$

63

$HO{-}CH(CH_3){-}CH_2CH_2{-}OH$

64

$HO{-}CH_2{-}CH{=}CH{-}CH_2{-}OH$

65

$H_3C{-}CH_2{-}CH{=}CH{-}CH_2CH_2{-}OH$

66

a: *Z*-isomer
b: *E*-isomer

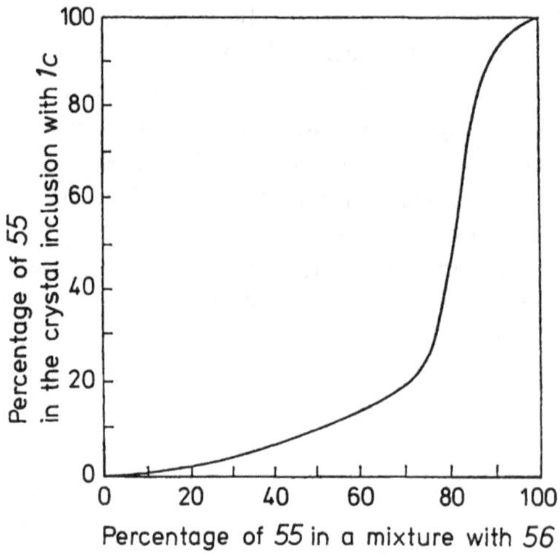

Fig. 9. Separation of *55* and *56* by inclusion formation with *1c* in benzene

interaction only since *51* is fixed in a crystalline lattice by the surrounding i-Pr groups of *23* [23]. 6-Acetyl-2-methylnaphthalene (*54*) is easily separated from the 1-acetyl isomer *53* by complex formation with *1a*. In this case, only the more symmetrical compound *54* forms a complex with *1a* [10].

Separation of 3-methylpyridine (*55*, bp 143.5 °C) from its 4-isomer (*56*, bp 143.1 °C) is also an important but very difficult problem in industry. However, since crystals of *1a* and *7* accommodate only *55* and *56*, respectively, these hosts are useful for separation of the picolines. A most interesting selective inclusion of the two pyridino candidates *55* and *56* by *1c* was also observed. As shown in Fig. 9, when the concentration of *55* exceeds the 85% level, an inclusion compound with *1c* is formed predominantly [10]. However, when the concentration of *55* remains under the 75% level, an inclusion compound is formed with *56* predominantly. If one uses this phenomenon skilfully, pure *55* and *56* can be isolated from any composition of a mixture of them. Separation of 2-methylquinoline (*57*) from its 8-isomer *58* can also be achieved by using *1c*, since only the former results in a crystal inclusion [10].

Cyclohexanone and cyclohexenone derivatives are important starting materials in organic synthesis. However, difficulties of purification of such compounds are sometimes encountered because they are often contaminated with isomers of almost the same boiling point. The crystal inclusion method is also applicable for the purification of these compounds. For example, since cis-3,5-dimethylcyclohexanone (*59*, bp 179 °C) forms an inclusion compound with *1a*, but not its trans-isomer *60* (bp 179 °C), they are easy to separate [10]. Similarly, *1c* forms a crystal inclusion with the conjugated bicyclic enone *61* (bp 143–145 °C/15 mm Hg) but not with the non-conjugated *62* (bp 143–145 °C/15 mm Hg) which allows separation [10].

The constitution of an alkyl chain is also well recognized by some hosts. For example, *24* recognizes the difference between the straight alkyl chain of 1,4-butanediol (*63*) and the branched chain of 1,3-butanediol (*64*), in that it includes the former selectivity. Regonition of geometrical isomerism around a double bond by a host compound seems to be quite easy. For example, *24* includes trans-2-butene-1,4-

diol (*65 b*) but not its cis-isomer *65 a* [10]. In the case of 3-hexen-1-ol (*66*), however, the *z*-isomer *66 a* is included in *24* selectively [10].

3.2 Isolation of Alcohols from Aqueous Solution

Isolation of water soluble materials from an aqueous solution is very costly and much problematic in an industrial scale. The isolation of ethanol from an aqueous solution obtained by the fermentation of biomass is one of the most important problems relating to a cheap energy source which may take the place of oil in future. If the crystal inclusion method can be applied to such an isolation process cheap ethanol is available which might cause a revolution in the utilization of energy. This Section deals with the isolation of ethanol and some other alcohols from aqueous solution.

We first found that *1 a* includes ethanol in the crystal loosely. The corresponding 1:1 inclusion compound loses ethanol at 84 °C but its 2,4-dimethyl-analogue *67 a* includes ethanol tightly and the resulting 1:2 crystal inclusion releases the two ethanols stepwise at 86 and 127 °C. The data suggested that the more voluminous 2,4-dimethylphenyl groups surround two ethanol molecules tightly but the smaller phenyl groups surround the ethanol only loosely in the crystal lattice of the corresponding inclusion compounds. The feature of the 1:2 ethanol inclusion of *67 a* is well shown in its X-ray crystal structure (Fig. 10) [24].

Interestingly, however, the 2-methyl-analogue *67 b* includes one ethanol loosely and releases it at 83 °C, but the 4-methyl- *67 c* and 2,5-dimethyl-analogues *67 d* do not accommodate ethanol in the crystal lattice. Furthermore, although *1 a*, *67 a*, and

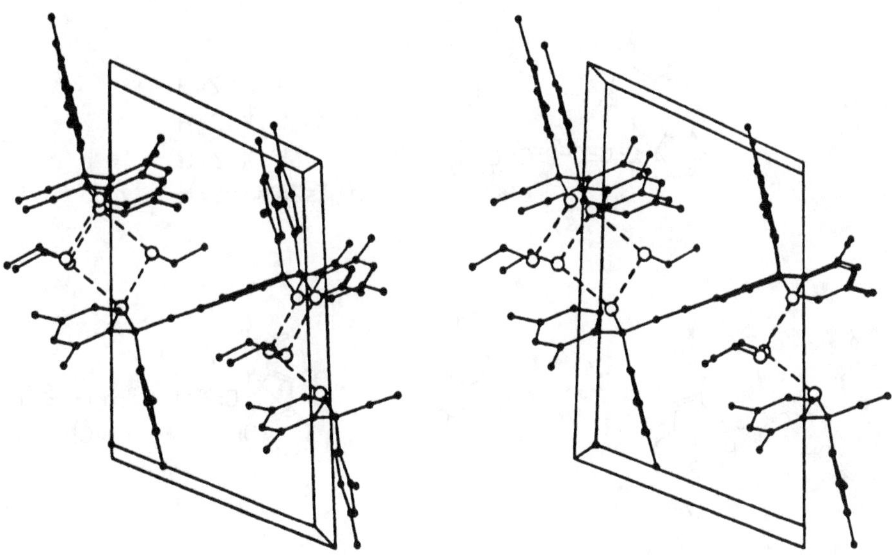

Fig. 10. Stereodrawing of the 1:2 ethanol inclusion compound of *67 a* (broken lines represent H-bond contacts; O atoms as circles, C atoms as bold dots)

Table 5. Some typical alcohols which are included in *1a*, *67a–d*, and *68a–b*, and molar ratios to the host compounds

Alcohol	Host compound						
	1a	*67a*	*67b*	*67c*	*67d*	*68a*	*68b*
Methanol	1	2	1	—	—	—	2
Ethanol	1	2	1	—	—	—	2
1-Propanol	—	—	—	—	—	—	—
2-Propanol	—	2	1	—	—	2	2
1-Butanol	—	—	—	—	—	—	2
Isobutanol	—	—	—	—	—	2	—
2-Butanol	—	—	1	—	—	—	—
t-Butanol	2	2	1	—	—	—	2

67b also include some other alcohols in the ratios shown in Table 5, *67c* and *67d* do not accommodate any kind of alcohol. It is also interesting that *67b* is effective for a greater variety of alcohols than does *67a* (Table 5). There might be a complicated steric factor of the host molecule which controls the stability of its crystal inclusion.

Higher inclusion ability of *68b* for alcohols as compared to *68a* is probably due to a similar steric factor (Table 5). Anyhow, refering to the above finding, efficient host compounds which have 2,4-dimethylphenyl groups (*17* and *69*) were designed for ethanol isolation.

The procedure of ethanol isolation from aqueous solution is very simple. For example, when a solution of *17* (26.4 g, 0.1 mol) in 50% ethanol (18.4 g, 0.2 mol

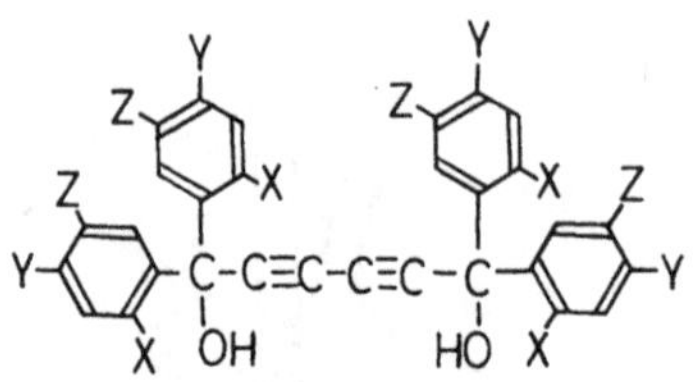

a: X=Y=Me; Z=H
b: X=Me; Y=Z=H
c: Y=Me; X=Z=H
d: X=Z=Me; Y=H

67

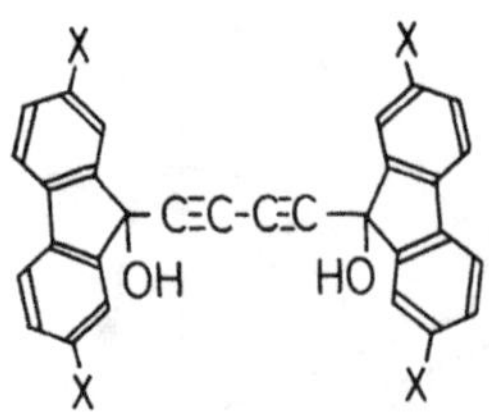

68

a: X=H
b: X=Cl

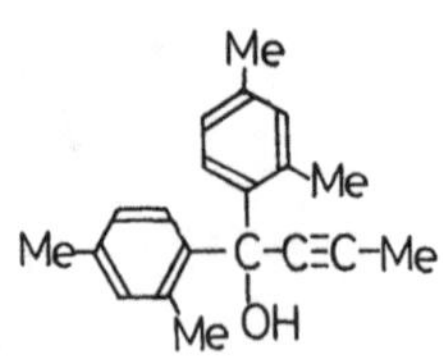

69

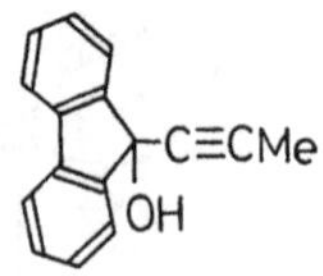

70

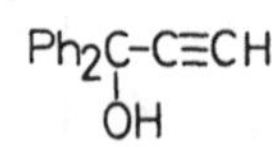

71

ethanol) is kept at room temperature for 3 h, a 1:1 ethanol inclusion compound of *17* is obtained as colorless crystals (30.4 g, 98% yield based on *17*). Heating of this inclusion compound gave 99% pure ethanol by distillation (4.3 g, 96% yield based on the crystal inclusion [10]. The free host is recovered and can be reused. Similarly, ethanol is isolated by inclusion formation with *69*. The X-ray crystal structure of the 1:1 ethanol inclusion compound of *69* has been reported [24]. It is assigned that the 1:1 ethanol inclusion of *17* has a crystal build-up analogous to that of *69*.

9-(1-Propynyl)-9-fluorenol (*70*), *5*, and *6b* were found to form stable crystal inclusions with ethanol and are also usable for ethanol isolation. For example, when a solution of *70* (10.0 g, 4.5 mmol) in 50% ethanol (100 ml, 850 mmol) was kept at room temperature for 12 h, a 1:1 ethanol inclusion compound of *70* was formed as colorless crystals (9.6 g, 80% yield based on *70*). Heating of the inclusion compound gave 98% pure ethanol (1.4 g, 85% yield based on the crystal inclusion [5]. X-Ray crystal structural studies of the ethanol inclusion compounds of *5* [6,8], *6b* [25] and *70*[24] have been reported.

Table 6. Some typical alcohols which are included in *11*, *17*, *69*, and *70*, and molar ratios to the host compounds

Alcohol	Host compound			
	11	*17*	*69*	*70*
Methanol	1	—	1	1
Ethanol	1	1	1	1
1-Propanol	1	—	1	1
2-Propanol	1	1	1	1
1-Butanol	1	—	—	—
Isobutanol	1	—	—	—
2-Butanol	1	—	1	—
t-Butanol	1	—	—	1

In addition to ethanol, a wide variety of other alcohols are included by the host compounds shown in Table 6. X-Ray crystal structures of the 1:2 methanol [8] and 1:1 butanediol inclusion compounds of *5* [7] and of the 1:1 butanediol crystal inclusion of *6c* [7] have been reported. By using inclusion crystal formation, it is possible to separate the corresponding alcohols from aqueous solution.

Amide host compounds are also useful for the isolation of alcohols. For example, when a solution of *24* (2.0 g) and 70% aqueous solution of 2-propyn-1-ol (0.73 g) in n-butanol was kept at room temperature for 1 h, a 1:2 inclusion compound of *24* and 2-propyn-1-ol was formed as colorless crystals (1.45 g), which upon heating at 150 °C gave pure 2-propyn-1-ol (0.27 g, 58%) [10]. By the same method, 1,4-butanediol, 2-butene-1,4-diol, 2-butyne-1,4-diol, ethylene glycol, and diethylene glycol were isolated in 50–70% yields from their aqueous solutions [10].

3.3 Isolation of Amines and Ammonia from Aqueous Solution

Many host compounds described in Section 2 form stable crystal inclusions with amines. This process can be applied for the storage and isolation of amines. For instance, when a solution of *7* (2.86 g, 10 mmol) and 50% aqueous Me_2NH (0.9 g, 10 mmol) in methanol (10 ml) was kept at room temperature for 1 h, a 1:1 Me_2NH inclusion compound of *7* was formed as colorless crystals (3.05 g, 92%, decomposes at 130–170 °C), which upon heating gave pure Me_2NH (0.41 g, 91%) [10]. By the same method, a 1:1 $MeNH_2$ crystal inclusion of *7* (decomposes at 140–190 °C) was obtained in 90% yield from a 70% aqueous $MeNH_2$ solution [10]. Analogous treatment of a 70% aqueous $EtNH_2$ solution with *7* in methanol gave a 1:1:1 crystal inclusion of *7* with $EtNH_2$, and methanol in 85% yield (decomposes at 100–170 °C) [10].

Some hosts also form inclusion compounds with NH_3. For example, a solution of *7* in NH_3-saturated methanol was kept at room temperature for 8 h. A 1:2 NH_3 inclusion compound of *7* was obtained as colorless crystals (decomposes at 80–120 °C) in a quantitative yield. Heating of it yields pure NH_3 quantitatively [10].

This type of inclusion formation is also unseful for the storage of volatile amines such as the above described $MeNH_2$ (bp —6.3 °C), $EtNH_2$ (16.6 °C), and Me_2NH (7 °C).

3.4 Isolation of Alkali Metal Hydroxide from Aqueous Solution

Unexpectedly, it was found that some phenol- and propynol-type hosts include alkali metal hydroxides (MOH) such as LiOH, NaOH, KOH, and CsOH, and form crystalline inclusion compounds. Phenol-type hosts, *7*, *8*, *11*, form crystal inclusions

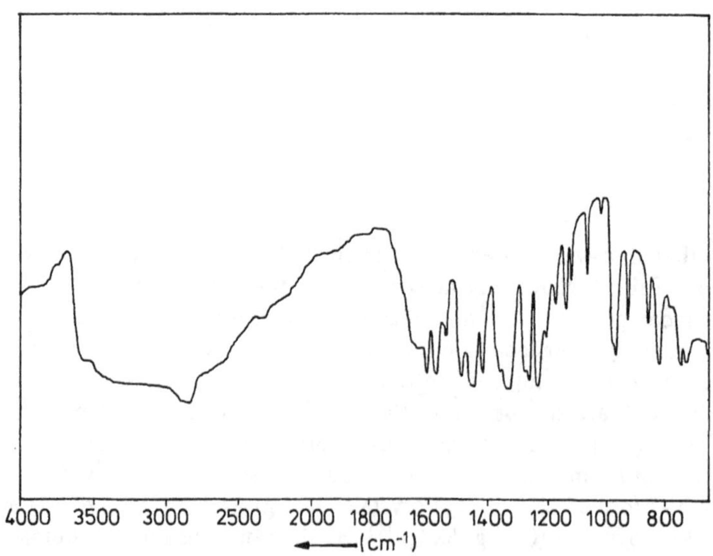

Fig. 11. IR spectrum of the NaOH crystal inclusion of *7* (7:NaOH:H_2O = 1.5:3:8) (Nujol mull)

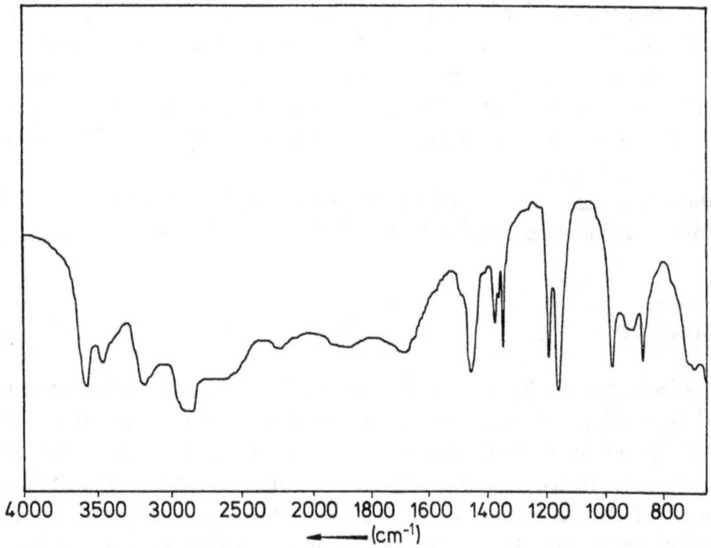

Fig. 12. IR spectrum of the NaOH crystal inclusion of *18* (*18*: NaOH: H_2O = 1:1:2) (Nujol mull)

of composition host: MOH: H_2O = 1:2–3:2–10. On the other hand, propynol-type hosts, *18* and *71*, form crystal inclusions of composition host: MOH: H_2O = 1:1:1–2. The composition varies depending on the kind of alkaline metal. The IR spectra of the NaOH inclusion compounds of *7* and *18* are shown in Fig. 11 and 12, respectively. The structures of this crystal inclusions are not yet clear, but strong hydrogen bonding between hydroxyl groups of the host and the hydroxide ions are very likely to play an important role for inclusion formation.

The finding of alkali metal hydroxide inclusions is important both from theoretical interest and from application. A possible consequence could be the isolation of alkali metal hydroxides from their aqueous solutions. Dried solid NaOH and KOH are made in industry by ordinary evaporation of their aqueous solutions obtained by electrolysis of NaCl and KCl, respectively. This method is uneconomical and wastes a lot of energy. On the contrary, the crystal inclusion method is very simple and economical Examples of such isolation processes are given below.

When a mixture of *71* (10 g, 75 mmol) and 30% NaOH aqueous solution (10 g, 75 mmol) is stirred at room temperature for 10 min, the NaOH inclusion of *71* is formed as colorless crystals. The complex can be dried in a desiccator and extracted with THF in a Soxhlet apparatus to give NaOH (2.58 g, 85%). Evaporation of the THF solution restores *71* (8.1 g, 81%). When a mixture of *18* (8.4 g, 0.1 mol) and a 30% aqueous solution (13.3 g, 0.1 mol) of NaOH was stirred at room temperature for 10 min, the NaOH inclusion of *18* was formed as colorless crystals which was dried in a desiccator. Heated under 1–2 mm Hg distills off *18* (7.1 g, 85%) to yield dried NaOH (3.4 g, 85%) as a residue. Analogous treatment of *18* (8.4 g, 0.1 mol) and a 50% KOH aqueous solution (11.2 g, 0.1 mol) gave dried (4.7 g, 84%) and the recovered host (7.0 g, 83%).

Phenol-type hosts also form stable crystal inclusions with alkali metal hydroxides. For example, when a mixture of *7* (28.6 g 0.1 mol) and a 5% LiOH aqueous solution was kept at room temperature for 6 h, a LiOH inclusion of *7* was obtained as colorless crystals (33.3 g, 90%) of composition *7*: LiOH:H_2O = 1.5:3:8. Analogous treatment of *7* (28.6 g, 0.1 mol) with 20% aqueous KOH (48.8 g, 92%) gave a KOH inclusion of *7* as colorless crystals (48.8 g, 92%) of composition *7*:KOH:H_2O = 1.5:3:8. Extraction of the host from the inclusion compounds with THF yields the pure alkali metal hydroxides.

Interestingly, these hosts also accommodate ammonium hydroxide (NH_4OH) and form stable crystalline inclusion compounds with it [10].

4 Optical Resolution

If host and guest molecules mutually recognize their chiralities at inclusion formation, the process could be used for optical resolution. In other words, when the host compound is optically active, one enantiomer of the guest compound should be included selectively. In turn, if an optical active guest molecule forms a crystal inclusion with one enantiomer of the host compound selectively, the host compound yields resolved. This section deals with the resolution via inclusion formation using host compounds such as alkaloids, 2-propyn-1-ols, 2,4-hexadiyne-1,6-diols, 2,2'-dihydroxy-1,1'-binaphthol (*7*), and 2,2'-dihydroxy-9,9'-spirobifluorene (*10a*).

4.1 Alkaloids

We found that some kinds of alcohols form channel-type inclusion compounds with alkaloids such as brucine (*72*) and sparteine (*73*), and that the alcohols are easily resolved utilizing inclusion formation. We also found that sparteine can be resolved by inclusion formation with an optically active tertiary acetylenic alcohol.

Table 7. Yields and $[\alpha]_D$ values of 100% ee *12a–m* obtained by optical resolution utilizing crystal inclusion with brucine

	12 R	Inclusion formation time	$[\alpha]_D$ (°) (c 1.0 in MeOH)	Yield (%)
a	o-Br—C_6H_4—	3	−134	63
b	o-Cl—C_6H_4—	8	−135	10
c	o-F—C_6H_4—	3	−59.1	14
d	o-Me—C_6H_4—	6	−53.7	21
e	t-Am	1	+10.5	84
f	t-Bu	2	+12.4	77
g	n-Bu	1	+7.9	47
h	n-Pr	2	+4.5	29
i	i-Pr	3	+1.1	19
j	Et	2	+7.2	36
k	CCl_3	4	+13.8	33
l	$CHCl_2$	2	−3.4	70
m	CH_2Cl	3	+11.1	39

Table 8. Yields and $[\alpha]_D$ values of 100% ee *13* and *14* obtained by optical resolution utilizing crystal inclusion with brucine

	Acetylenic alcohol R	Inclusion formation time	$[\alpha]_D$ (°) (c 0.1 in MeOH)	Yield (%)
13	Ph	5	+116	10
13	Et	2	+55.4	38
13	CH_2Cl	3	+58.1	20
14	n-Am	2	+5.5	53
14	n-Bu	2	+3.1	61
14	n-Pr	2	+3.9	37
14	i-Pr	2	+4.5	45

For example, when a solution of *12f* (8.12 g, 43.2 mmol) and *72* (17.0 g, 43.2 mmol) in acetone (260 ml) was kept at room temperature for 12 h, a 1:1 crystal inclusion of (+)-*12f* and *72* (12.1 g) was obtained as colorless crystals. Decomposition of the inclusion compound with dil. HCl gave 71% ee of (+)-*12f* (3.9 g, 96%), which upon a second crystallization with *72* was raised to 100% ee of (+)-*12f* (3.16 g, 77%, $[\alpha]_D$ +12.4°). The acetone solution left after separation of the brucine inclusion

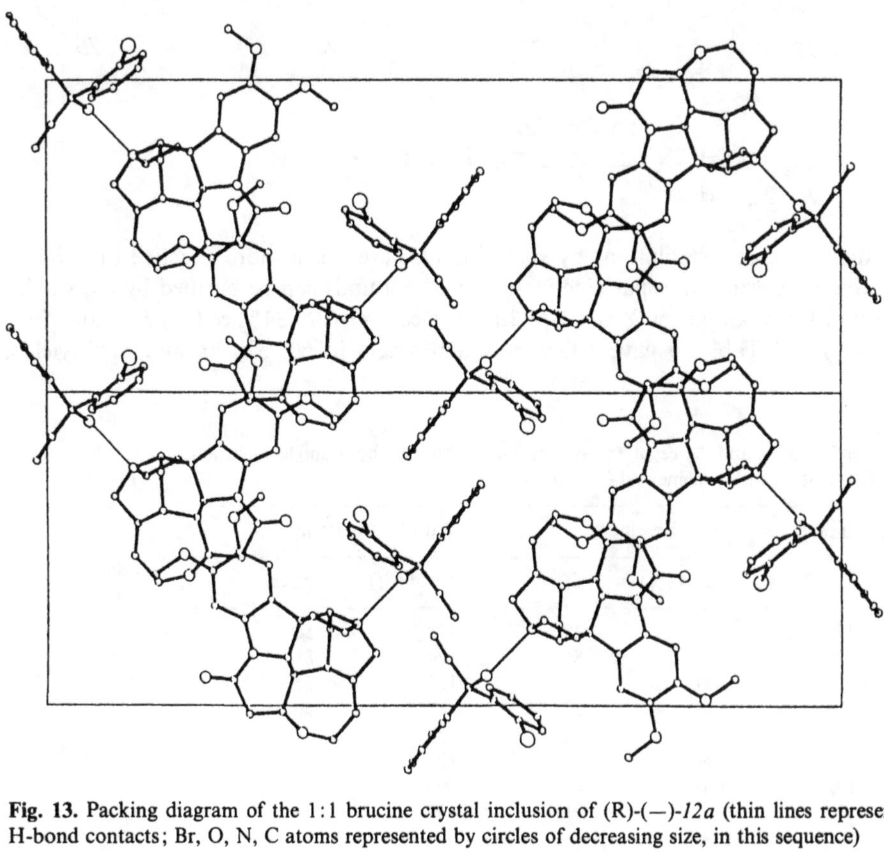

Fig. 13. Packing diagram of the 1:1 brucine crystal inclusion of (R)-(−)-*12a* (thin lines represent H-bond contacts; Br, O, N, C atoms represented by circles of decreasing size, in this sequence)

compound of 71% ee of (+)-*12f*, was treated as above to give 66% ee of (—)-*12f* (4.06 g, 100%) [11,13]. By the same procedure, *12a–e* and *12g–m* were also resolved efficiently to give 100% ee enantiomers finally (Table 7) [13]. In Table 7 only the yields of the guest enantiomers which crystallize out as inclusion compounds with *72* are given. Since *12* with R=Me or CF_3 is only poorly resolved by *72*, R should be a larger group than Me to make an efficient resolution possible. Like *12*, *13* and *14* were also easily resolved giving 100% ee enantiomers (Table 8) [13].

This effective optical resolution is possible because the enantiomers mirror-related to those listed in Tables 7 and 8 do not allow inclusion formation with *72* and *73*. The X-ray crystal structural study of the 1:1 inclusion compound of (R)-(—)-*12a* and *72* shows that the host enantiomer is accomodated in the channel space formed by *72*, and that two kinds of hydrogen bonds, OH ··· N of *72* and C≡CH ··· O of *12a*, play an important role to hold the guest components together (Fig. *13*). Replacement of the (R)-(—)-*12a* molecules in this lattice aggregate by the other guest enantiomer, (S)-(+)-*12a*, would result in an unfavorable system of hydrogen bonds.

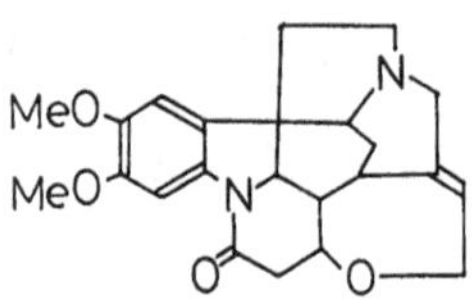

72

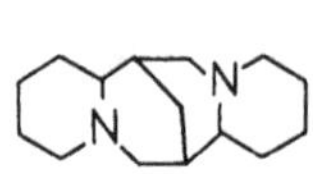

73

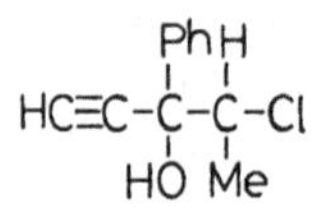

74

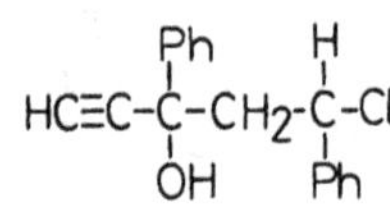

75

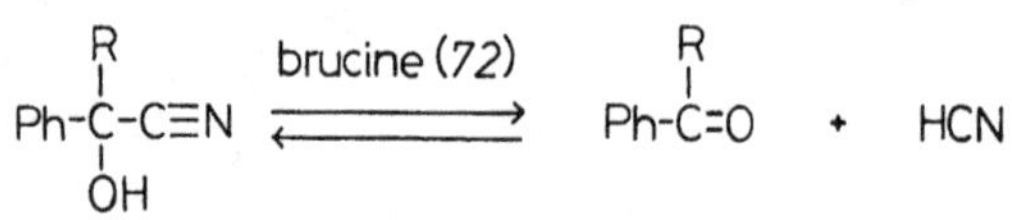

(Eq. 1)

In some cases, resolutions by sparteine (*73*) are much more effective than by *72* (Table 9), because the sparteine inclusion compounds can be purified by recrystallization. For example, 50% ee (—)-*12a*, 55% ee (—)-*12b*, 34% ee (—)-*12c*, and 59% ee (—)-*12f* (Table 9) gave 100% ee enantiomers in 60, 52, 20, and 62% yields,

Table 9. Yields and % ee of the enantiomers obtained by a single inclusion crystallization with brucine and (—)-sparteine

Enantiomer	With brucine		With (—)-sparteine	
	Yield (%)	% ee	Yield (%)	% ee
(—)-*12a*	130	39	122	50
(+)-*12a*	70	81	74	81
(—)-*12b*	120	15	102	55
(+)-*12b*	76	22	88	60
(—)-*12c*	90	60	72	34
(+)-*12c*	104	53	122	23
(—)-*12f*	100	66	98	59
(+)-*12f*	96	71	102	57

respectively, by two recrystallizations of their 1:1 sparteine inclusions from acetone followed by decomposition. On the other hand, racemic sparteine was easily resolved by utilizing the inclusion formation with an optically active propynol [12]. This method can probably be applied to the resolution of many synthetic alkaloids and related compounds.

Propynols which have two chiral carbons also form inclusion compounds with *72*, and both chiral centers can be resolved quite efficiently by inclusion formation. For example, the 1,2- and 1,3-double chiral centers of *74* and *75*, respectively, are resolved perfectly by inclusion formation with *72*, and gave 100% optically pure compounds [26].

This resolution method was found to be applicable to some cyanohydrins *15* and secondary alcohols *16*. Very interestingly, the racemic cyanohydrins were converted into a pure optically active isomer in almost quantitative yield in the presence of *72* [27]. For example, when a solution of racemic 1-cyano-2,2-dimethyl-1-phenylpropanol (*15a*) (1.0 g, 5.3 mmol) and *72* (2.1 g, 5.3 mmol) in methanol (2 ml) was kept in an uncapped flask for 24 h at room temperature, a brucine inclusion compound of 94% ee of (+)-*15a* was obtained in quantitative yield, which upon decomposition gave 94% ee of (+)-*15a* (1.0 g). Repeating inclusion formation of the 94% ee containing (+)-*15a* (1.0 g) with *72* (2.1 g) one more time yielded 100% ee of (+)-*15a* [1.0 g, $[\alpha]_D$ + 15.9° (c 1.0 in MeOH)]. The process of the complete conversion of racemic cyanohydrin to one enantiomer consists of racemization of cyanohydrin through the equilibrium in Equation 1 and selective inclusion of one enantiomer in brucine [27].

Simple secondary alcohols *16* are also easily resolved with brucine to give 100% ee of *16a* [$[\alpha]_D$ 32.3° (c 1.0 in MeOH)] and *16b* [$[\alpha[_D$ 34.4° (c 1.0 in MeOH)] [10]. However, in the cases of *15* and *16*, a sterically bulky group such as t-Bu or CCl_3 is necessary for efficient optical resolution.

4.2 1,6-Diphenyl-1,6-di(o-chlorophenyl)hexa-2,4-diyne-1,6-diol

Since *1a* is a good host molecule for including a wide variety of organic compounds, an optically active hexadiynediol such as the title compound *76* is suggested to be a good host for optical resolution of different guest compounds. Oxidative coupling of optically pure *12b* prepared according to the resolution method of Section 4.1, easily led to optically pure *76* [$[\alpha]_D$ 122° (c 1.0 in MeOH)].

2,3-Epoxycyclohexanones *77–79* were easily resolved by *76* [28]. For example, when a solution of (—)-*76* (5.10 g, 10.6 mmol) and *77* (2.97 g, 21.2 mmol) in 1:1 ether light petroleum (20 ml) was kept at room temperature for 6 h, a 1:1 inclusion compound of (—)-*76* with (—)-*77* (4.68 g, 71%) was obtained as colorless crystals, which upon Kugelrohr distillation in vacuo gave 90% ee of (—)-*77* (70%). One recrystallization of the 1:1 inclusion of (—)-*76* with 90% ee (—)-*77* (4.68 g) from 1:1 ether light petroleum (100 ml) gave the inclusion of (—)-*76* with 100% ee (—)-*77* (3.50 g, 53%), which upon Kugelrohr distillation in vacuo gave 100% ee of (—)-*77* [0.6 g, 40%, $[\alpha]_D$ —136° (c 1.0 in MeOH)]. By the same method, *78* and *79* were also resolved to give 100% ee of (+)-*78* [31%, $[\alpha]_D$ +13.5° (c 1.0 in MeOH)] and of (+)-*79* [18%, $[\alpha]_D$ +58.3° (c 1.0 in MeOH)], respectively. The host *76* left after distilling off *77–79* can be used again.

To our knowledge, no such effective optical resolution by inclusion formation with an optically active channel-type host compound has been reported so far. 1,2-Bis(2-methyl-1-naphthyl)-1,2-bis(2,4,6-trimethylphenyl)ethane (*80*) yields a 1:1 crystal inclusion with (+)-α-pinene, but successive recrystallization of *80* from (+)-α-pinene gives only a partial optical resolution [29].

The above resolution method can also be applied to 3-methylcyclohexanone (*81*) and 5-methyl-γ-butyrolactone (*82*) yielding 100% ee of (+)-*81* [13%, $[\alpha]_D$ +141.4° (c 1.0 in MeOH)] and of (+)-*82* [4.5, $[\alpha]_D$ +30.1° (c 1.0 in MeOH)] [30]. Application of the method to resolve some key intermediates of the synthesis of prostaglandins and related compounds is also successful. For example, optically active 6-oxabicyclo-[3.3.0]oct-2-en-7-one (*83*) and 7-oxabicyclo[4.3.0]non-2-en-8-one (*85*) were readily obtained by the resolution method in good yields [31]. It is also easy to attain resolution of the dihydro derivatives of *83* (*84*) and *85* (*86a*) as well as of *cis*-bicyclo[4.4.0]deca-3-one (*87a*) and 6-methylbicyclo[4.4.0]deca-1-en-3-one (*88b*) [31].

4-Hydroxy-2-cyclopentenone (*89*) which is another important starting material for the prostaglandin synthesis is very difficult to obtain in optically pure form. The inclusion method can again be applied very successfully to the resolution of the esters of *89* (*90a–c*). For example, when a solution of (—)-*76* (4.83 g) and racemic *90a* (3.36 g) in 1:2 ether light petroleum (15 ml) was kept at room temperature for 12 h, a 1:1 inclusion compound of (—)-*76* and (—)-*90a* (4.69 g) was obtained as colorless crystals. The compound was purified by recrystallization from 1:2 ether light petroleum and heated in vacuo to give 90% ee of (—)-*90a* [0.94 g, 56%, $[\alpha]_D$ —134° (c 1.0 in MeOH)]. The filtrate was treated with (+)-*76* as above to give finally 100%

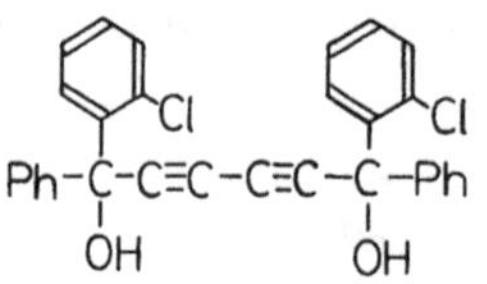

76

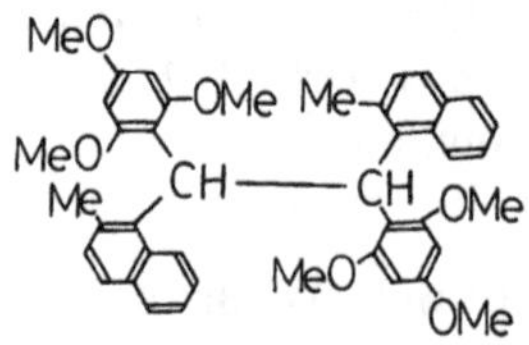

80

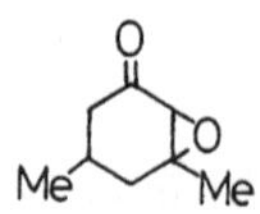
77

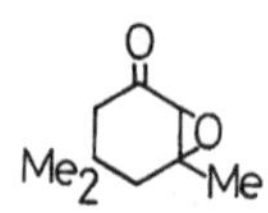
78

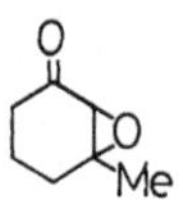
79

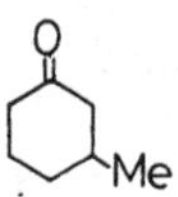
81

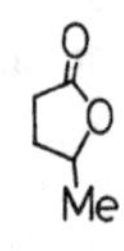
82

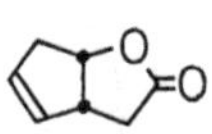
83

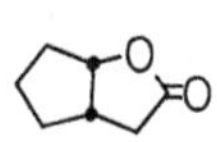
84

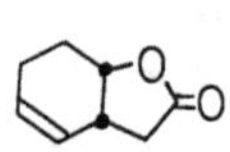
85

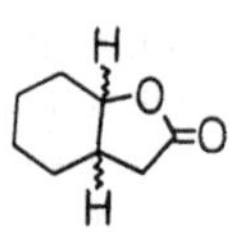
86

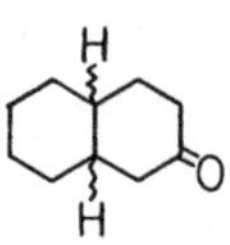
87

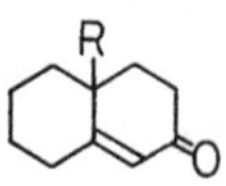
88

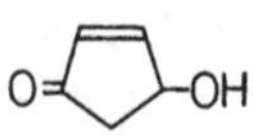
89

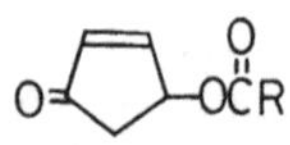
90

a: cis-isomer
b: trans-isomer

a: R = H
b: R = Me
c: R = Et

a: R = n-Pr
b: R = n-Bu
c: R = t-Bu

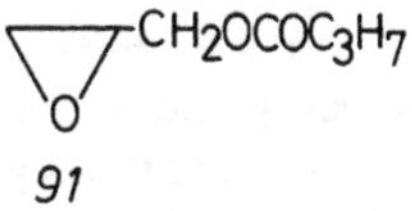
91

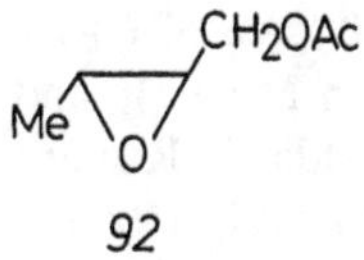
92

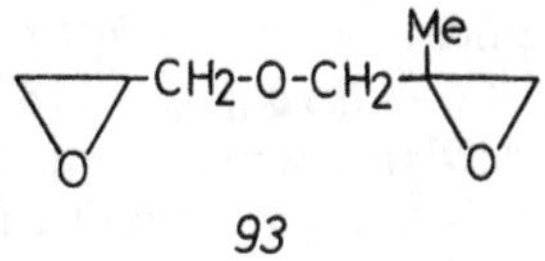
93

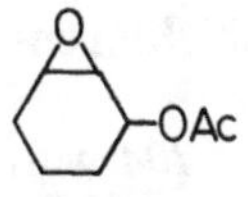
94

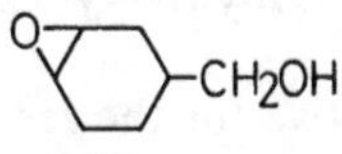
95

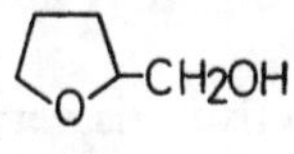
96

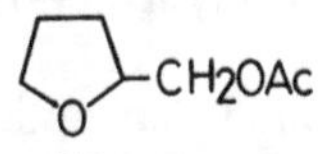
97

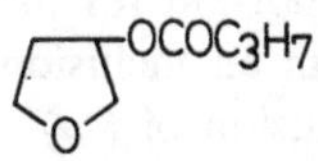
98

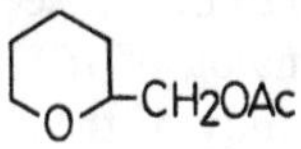
99

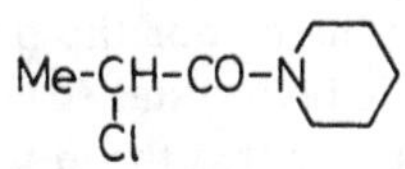
100

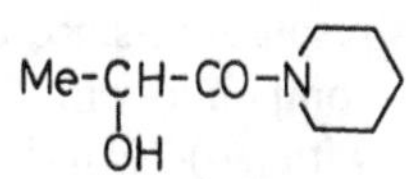
101

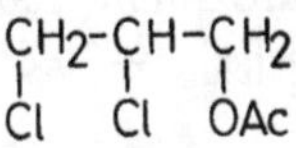
102

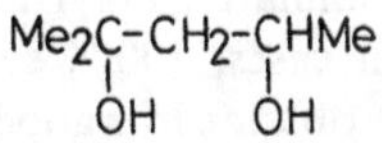
103

ee of (+)-*90a* in 56% yield [10]. By the same method, *90b* and *90c* are also easily resolved. Hydrolysis of the resolved *90a–c* yields optically pure *89* [10].

Cyclic ethers such as *91–99* are also very effectively resolved by (—)-*76* to give the optically pure enantiomers: (—)-*91* [8%, $[\alpha]_D$ —11.8° (c 1.0 in MeOH)], (+)-*92* (31%, +50.6°), (—)-*93* (22%, —20.2°), (—)-*94* (18%, —98.1°), (—)-*95* (52%, —8.8°), (+)-*96* (15%, +14.9°), (+)-*97* (56%, +27.1°). (+)-*98* (42%, +13.7°), and (—)-*99* (30%, —6.5°), respectively.

Acyclic compounds can also be resolved by inclusion formation with *76*, e.g. *100–103*. They are very effectively resolved by (—)-*76* to give the optically pure enantiomers: (+)-*100* [40%, $[\alpha]_D$ +23.5° (c 1.0 in MeOH)], (—)-*101* (64%, —2.3°), (+)-*102* (51%, +18.6°), and (—)-*103* (25%, —9.6°).

Some alkyl methyl sulfoxides *104* which have a relatively long alkyl chain are easily resolved by complexation with *76*. For example, *104a*, *104b*, and *104c* were

104

a: n = 6 *d*: n = 3
b: n = 5 *e*: n = 2
c: n = 4

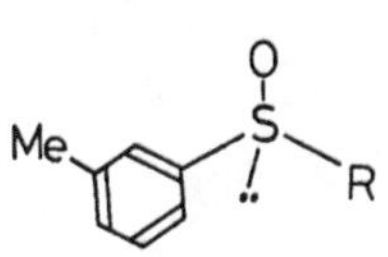
105

a: R = Me
b: R = Et
c: R = CH=CH$_2$

resolved efficiently by (—)-*76* to give 100% ee of (+)-*104a* [[α]$_D$ +105° (c 1.0 in MeOH)], of (+)-*104b* [[α]$_D$ +118° (c 1.0 in MeOH)], and of (+)-*104c* [[α]$_D$ +99.5° (c 1.0 in MeOH)], respectively, in good yields [10]. Remarkably, however, the resolution by *76* is not very effective for *104d* and *104e* having relatively short alkyl chains, and also for *104* — type compounds containing a branched alkyl group.

4.3 2,2-Dihydroxy-1,1′-binaphthyl and 2,2′-Dihydroxy-9,9′-spirobifluorene

It was found that optically active *7* is very effective for the resolution of *104* with relatively short alkyl chains such as *104d–e* and for m-tolyl alkyl sulfoxides *105*. In these cases, 100% ee of (+)-*104d* [[α]$_D$ +111° (c 1.0 in MeOH)], of (+)-*104e* (+123°), of (+)-*105a* (+140°), of (+)-*105b* (+199°), and of (+)-*105c* (+486°) were obtained in good yields [32]. The m-tolyl group of *105* seems important for the resolution, since neither the o- nor the p-tolyl-analog of *105* do form an inclusion compound with *7*. The X-ray crystal structural study of the 1:1 inclusion of (+)-*7* with (+)-*105a* (Fig. 6) shows that the m-tolyl group of *105a* is nicely accommodated in the host lattice of the inclusion compound [9].

Utilizing this unique chiral recognition between *7* and sulfoxides, racemic *7* was resolved very efficiently by (+)-*105a* [32]. Similarly, alkyl m-tolyl sulfoximine *106* was also easily resolved by (+)-*7*. In both cases, 100% ee of *106a* [[α]$_D$ +30.8° (c 0.5 in THF)] and of *106b* (+28.0°) were obtained in good yields [10].

A similar twisted host compound *10a* can be used for optical resolution instead of *7*. For example, racemic *10a* was resolved by inclusion formation with sparteine (*73*) very efficiently to give 100% ee of the (+)-*10a* enantiomer [[α]$_D$ +26.6° (c 1.0 in MeOH)]. Reversely, racemic *73* was resolved by optically active *10a*. Optically active *10a* is also useful for resolution of guest compounds such as 3-methylpiperidine (*107*) [10].

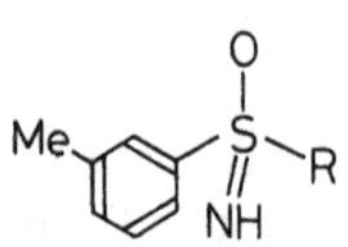

106

a: R = Me

b: R = Et

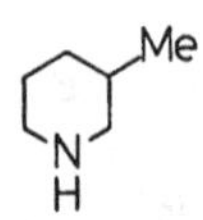

107

5 Acknowledgements

The author is grateful for the many contributions of his coworkers whose names appear in the references. The author would especially like to thank Professor Thomas C. W. Mak (a contribution of this author is on page 141) for his valuable X-ray analyses of many of the inclusion compounds and Professor Harold Hart for his helpful suggestions and encouragement.

6 References

1. Toda, F., Akagi, K.: Tetrahedron Lett. 3695 (1968)
2. Toda, F., Tanaka, K., Hart, H., Ward, D. C., Ueda, H., Ōshima, T.: Nippon Kagaku Kaishi 239 (1983)
3. Toda, F., Ward, D. L., Hart, H.: Tetrahedron Lett. *22*, 3865 (1981)
4. Toda, F., Tanaka, K., Wang, Y., Lee, G.-H.: Chem. Lett. 109 (1986)
5. Toda, F., Tanaka, K., Ulibarri Daumas, G., Sanchez, M. C.: ibid. 1521 (1983)
6. Toda, F., Tanaka, K., Nagamatsu, S.: Tetrahedron Lett. *25*, 1359 (1984)
7. Toda, F., Tanaka, K., Mak, T. C. W.: J. Incl. Phenom. *3*, 225 (1985)
8. Toda, F., Tanaka, K., Nagamatsu, S., Mak, T. C. W.: Isr. J. Chem. *25*, 346 (1985)
9. Toda, F., Tanaka, K., Mak, T. C. W.: Chem. Lett. 2085 (1984)
10. Toda, F. et al.: unpublished data
11. Toda, F., Tanaka, K., Ueda, H.: Tetrahedron Lett. *22*, 4669 (1983)
12. Toda, F., Tanaka, K., Ueda, H., Ōshima, T.: J. Chem. Soc., Chem. Commun. 743 (1983)
13. Toda, F., Tanaka, K., Ueda, H., Ōshima, T.: Isr. J. Chem. *25*, 338 (1985)
14. Recca, A., Bottino, F. A., Libertini, E., Finocchiaro, P.: Gazz. Chim. Ital. *109*, 213 (1979)
15. Din, L. B., Meth-Cohn, O.: J. Chem. Soc., Chem. Commun. 741 (1977)
16. MacNicol, D. D., Wilson, D. R.: Chem. & Ind. (London) 84 (1977)
17. Ichikawa, Y., Tsuruta, H., Kato, K., Yamanaka, Y., Yamamoto, A.: Japanese Patent, 65,859 (1976)
18. Suzuki, Y., Maki, T., Mineta, K.: Japanese Patent, 96,530 (1975)
19. Brown, D. H., Cross, R. J., Mallinson, P. R., MacNicol, D. D.: J. Chem. Soc., Perkin Trans. 2, 993 (1980)
20. Allcock, H. R., Allen, R. W., Bissell, E. C., Smelts, L. A., Teeter, M.: J. Am. Chem. Soc. *98*, 5120 (1976)
21. Allcock, H. R., Stein, M. T.: ibid. *96*, 49 (1974)
22. Allcock, H. R., Siegel, L. A.: ibid. *98*, 5140 (1976)
23. Toda, F., Tanaka, K., Tagami, Y., Mak, T. C. W.: Chem. Lett. 195 (1985)
24. Toda, F., Tanaka, K., Mak, T. C. W.: Bull. Chem. Soc. Jpn. *58*, 2221 (1985)
25. Toda, F., Tanaka, K., Mak, T. C. W.: Chem. Lett. 1699 (1983)
26. Toda, F., Tanaka, K., Mori, K.: ibid. 827 (1983)
27. Toda, F., Tanaka, K.: ibid. 661 (1983)
28. Tanaka, K., Toda, F.: J. Chem. Soc., Chem. Commun. 1513 (1983)
29. Hayes, K. S., Hounshell, W. D., Finocchiro, P., Mislow, K.: J. Am. Chem. Soc. *99*, 4152 (1977)
30. Toda, F., Tanaka, K., Nakamura, K., Ueda, H., Ōshima, T.: ibid. *105*, 5151 (1983)
31. Toda, F., Tanaka, K.: Chem. Lett. 885 (1985)
32. Toda, F., Tanaka, K., Nagamatsu, S.: Tetrahedron Lett. *25*, 5929 (1984)

Tri-o-Thymotide Clathrates

Raymond Gerdil

Département de Chimie Organique, Université de Genève,
30 quai Ernest-Ansermet, 1211 Genève, Switzerland

Table of Contents

1 Introduction

A remarkable property of the chiral and flexible tri-o-thymotide molecules (in the following designated as TOT; see Fig. 1) is their propensity to enclose a wide variety of guest molecules on crystallization. Actually it is rather difficult to find a simple compound which would not form a clathrate with TOT (not far from 100 clathrates are mentioned in the literature). As a consequence, the type and the nature of host-guest interactions subject to investigation are considerably broadened and diversified.

In the past ten to fifteen years there has been a strong recrudescence of interest in chiral phenomena, especially in the synthesis and development of novel chiral systems in solution. A similar trend is making great strides in the chemistry of the organic solid-state where new modes of creating and studying interacting chiral systems are in rapid progress [1]. In this respect the crystal chemistry of the clathrates of TOT, which originated some 30 years ago [2], provides numerous possibilities of investigating weak chiral molecular interactions in crystalline aggregates. These fortunate circumstances rely upon the property of most of TOT clathrates to undergo spontaneous resolution on crystallization as opposed to the vast majority of other chiral host molecules which afford exclusively achiral clathrate crystals.

1.1 Molecular Trigonal Symmetry and Host Properties

The constitution of TOT implies trigonal symmetry for the individual molecule. In all of its clathrates the three-propeller-shaped TOT molecule (Fig. 1) is constrained to adopt an asymmetrical conformation owing to the action of crystal packing forces.

MacNicol et al. [3] have stressed the fact that trigonal symmetry is a constitutional feature common to several important hosts. A notable fact is that trigonal (or hexagonal) lattice symmetry is often preserved in spite of local conformational distortion of the component host molecules. On the basis of relevant structural analogies in relation with the preceding symmetry considerations it has been

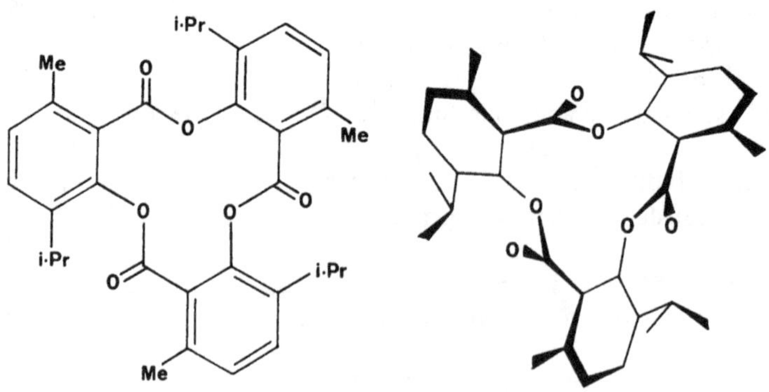

Fig. 1. Tri-o-thymotide. Constitution and idealized view of the (M)-(—)-configuration

possible to design and synthetize closely related trigonal and hexagonal host molecules [4]. On the contrary, the inclusion-forming properties of TOT seem to be intrinsically associated with the nature of the alkyl substituents and their location on the trisalicylide backbone, as was already illustrated by an earlier investigation [5]. For the present, N,N'-dimethyl-N''-benzyltri-3-methyltrianthranilide is the only known 12-membered ring akin to TOT to undergo spontaneous resolution and also form inclusion compounds [6].

1.2 Recent Chemistry of TOT

Until recently, the only reported product of the condensation of o-thymotic acid *1* by phosphoric anhydride or phosphorous oxychloride was a mixture of di-o-thymotide (DOT) and tri-o-thymotide in moderate yield. In order to improve the yield of TOT a careful reinvestigation of the reaction with $POCl_3$ or dicyclohexylcarbodiimide as condensation agents was carried out by Arad-Yellin et al. [7]. Several other esters were indeed isolated and characterized (*2*, *3*, *4*). It was suggested that these products are competitively generated by the facile decarboxylation of the intermediate compound *5* on the reaction path to intramolecular esterification. Good evidence was obtained that the hydroxy group assists decarboxylation of *5*.

Farias and Hosangadi [8] had previously studied the selective action of polyphosphate ester on the condensation reaction of o-thymotic acid. The process is characterized by the selective formation of the eight-membered di-o-thymotide at the expense of TOT. Mechanistic aspects for this specific cyclisation reaction are unknown at the present time.

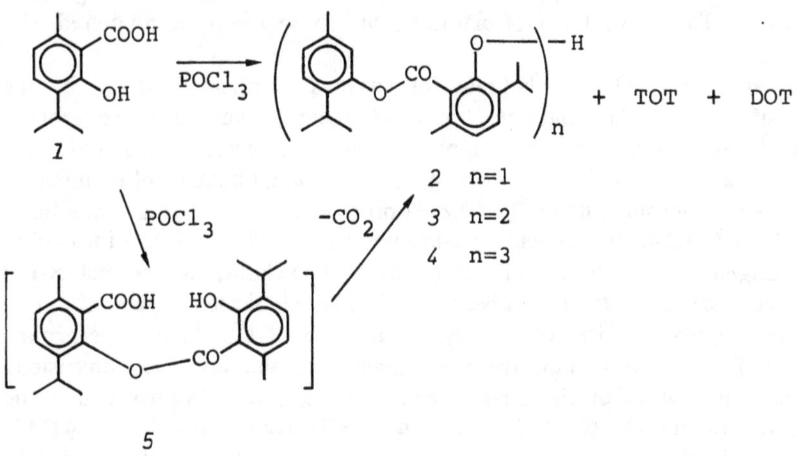

2 Structure of TOT Inclusion Compounds

TOT forms exclusively crystalline multimolecular inclusion compounds. The present section is mainly concerned with the stereochemical aspects of the following two types of TOT clathrates:

a) The guest molecules are enclosed in discrete closed cavities (cages) and are not in mutual contact. This is the most extensively studied type of TOT clathrates.

b) The guest molecules are accommodated in continuous "linear" cavities (channels) running through the crystal along characteristic crystallographic directions. The guest components are generally, but not always, in contact with each others in the channels.

2.1 Crystallographic Investigations

X-ray diffraction affords the basic method for investigating structural relationships in crystalline clathrates. Despite the large number of TOT clathrates mentioned in the literature, relatively few crystal structures have been completed. This situation is probably not fortuitous: according to the current interest in chiral phenomena the attention has been mainly focused on enantiomorphous TOT clathrates which, within each of two distinct types, are isostructural. This feature might have detered further X-ray analyses because of the ensuing restricted prospect of making new structural discoveries.

2.1.1 Crystal Data

In their earlier investigation Lawton and Powell [2b] determined the unit cell dimensions of about 60 chiral TOT clathrates pertaining to either of two distinct enantiomorphous types:

a) cage-type clathrates, space group $P3_121$, formed with guests of length not greater than about 9 Å;

b) channel-type clathrates, space groups $P6_1$, $P6_2$, $P3_1$, formed with long chain-like molecules. These two types of clathrates are those mostly encountered with TOT.

The versatility of TOT in clathrate formation is further illustrated by the discovery of several other clathrate types belonging to various space groups: $P\bar{1}$ (*trans*-stilbene, *cis*-stilbene [9], α-bromobutyric acid [10], benzene [11]); $P1(P\bar{1})$ (methyl *trans*-cinnamate, methyl *cis*-cinnamate); $P2_1$ (*meso*-2,3-butanediol carbonate); $P2_1/c$ (*meso*-2,3-di-bromobutane) [9]; $C2/c$ (3-bromooctane [9], ethyl α-bromobutyrate [10]); *Pbca* (1,1,1-trifluoro-2-chloro-2-bromoethane); *Pbcn* (dl-2,3-dibromobutane) [9], hexagonal R (α-chlorotetrahydropyran) [12]. The clathrates for which X-ray crystal structure determinations have been completed are indicated in Table 1.

Owing to the isomorphism of the crystals in each of the three space groups reported in Table 1, a preliminary measurement of the unit cell dimensions gives a good indication of the type of clathrate dealt with. Values within the following ranges are those mostly observed: $P3_121$ (cages) $a = b = 13.4–13.7$, $c = 30.0–30.6$ Å; $P6_1$ (channels parallel to c-axis) $a = b = 14.20–14.35$, $c = 29.0$ to 29.2 Å; $P\bar{1}$ (multi directional channels of variable sections) $a = 11.3–11.6$, $b = 13.0$ to 13.2, $c = 24.2–25.0$ Å, $\alpha = 93–96$, $\beta = 102–104$, $\gamma = 83–86°$.

2.1.2 Absolute Configurations of TOT

The existence of two pairs of enantiomeric TOT conformations in solution has been early detected by NMR spectroscopy [17]. The major form was referred to as

Table 1. Guest molecules in TOT clathrates structures determined by X-ray diffraction.

Space group	Guest[a]	R[b]	Ref.
$P3_121$	(dl)-2-bromobutane[c]	0.081	13b)
	(S)-(+)-2-bromobutane[d]	0.076	9a, 14)
	(R)-(—)-2-butanol[c, e]	0.068	13a)
	chlorocyclohexane[c]	0.061	12)
	(RR)-(+)-*trans*-2,3-dimethyloxirane[e]	0.064	9a)
	(dl)-*trans*-2,3-dimethylthiirane	0.087	9a)
	(SS)-(—)-*trans*-2,3-dimethylthiirane[e]	0.083	9a, 14)
	ethanol	—	15)
	(dl)-ethyl methyl sulfoxide[c]	0.049	13c)
	pyridine[c]	0.08	16a)
$P6_1$	n-cetyl alcohol	—	15)
$P\bar{1}$	benzene[c]	0.078	11)
	cis-stilbene[c]	0.21	9b)
	trans-stilbene[c]	0.131	9b)
$Pna2_1$	unsolvated TOT[c]	0.050	16b)

[a] Host:guest ratio is 2:1, with the exception of TOT/benzene being 2:2.5. [b] Residual index. [c] Atomic coordinates available from the Cambridge Crystallographic Database (Allen, F. H. et al: Acta Cryst. *B35*, 2331 (1979)). [d] Grown from enantiomerically enriched solution, purity of guest in crystal estimated to be 70%. [e] Optically pure

the propeller conformation and has all three carbonyl groups pointing to one side of the central twelve-membered ring (Fig. 1). The minor form was referred to as the helical conformation and has one carbonyl pointing in a direction opposite to that of the other two. Only the propeller form has been observed in the TOT clathrates studied so far. However, the surprising X-ray structure of tri-o-carvacrotide [18], a constitutional isomer of TOT formed by the interchange of the positions of the methyl and isopropyl substituents, shows that the racemic cristal lattice consists of a (1:1) population of propeller and helical chiral conformations. Two other crystal structures of tri-thiosalicylide derivatives illustrate further the geometry of the helix-like conformation [18].

The answer to the question posed by the absolute configuration of TOT is fundamental with a view to carrying out a correlation of configurations. Recent crystallographic results contradict a former configuration assignement by circular dichroism measurements [19]. Gerdil and Allemand obtained the absolute configuration of TOT from the clathrate formed with optically pure (R)-(—)-2-butanol [20]. This clathrate lacks enantioselectivity and diastereomeric pairs such as (—)-TOT/(R)-2-butanol and (+)-TOT/(R)-2-butanol have been isolated as single crystals. The crystallographic model arrived at was (M)-(—)-TOT/(R)-2-butanol (Fig. 2). To give stronger support to this model five additional single crystals of known optical activity were measured at 123 K. The structure factors were calculated for both diastereomeric models; subsequent tests of the *R*-factor ratio [21] confirmed the (M)-configuration for (—)-TOT. A similar conclusion [14] was reached independently with TOT clathrate crystals grown from optically enriched (S)-(+)-2-bromobutane and optically pure (R,R)-(+)-*trans*-2,3-dimethylthiirane, respectively (see Table 1), thus corroborating unequivocally the absolute configuration of TOT.

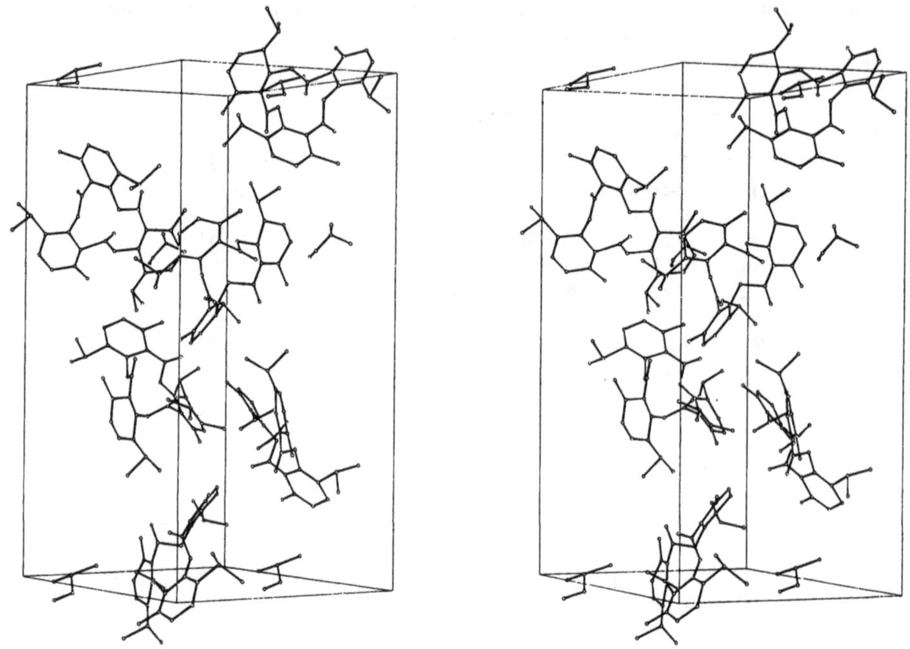

Fig. 2. Stereoview of the molecular packing of (M)-(—)-tri-o-thymotide with (R)-(—)-2-butanol as guest molecule. The unit cell is shown with the *c* axis pointing upward

2.1.3 Conformational Changes of TOT

The constitutional C_3 symmetry of the individual TOT molecule is not retained in the crystal lattice owing to packing forces exerted upon the molecule located in general position. It seems however that characteristic conformational deviations from the ideal C_3 symmetry are exhibited in each of the space groups so far investigated. Several lines of approach have been used to picture these deviations.

Williams and Lawton [15] have presented a graphical comparison of the equivalent torsion angles in the 12-membered central ring of TOT for three different crystalline species: the ethanol and n-cetyl alcohol TOT inclusion compounds, and the unsolvated TOT. This comparison was instructive because it revealed a fairly consistent molecular deformation in three different crystalline environments. A rationale for this structural feature is not readily apparent.

An alternative description of the conformational changes of TOT in various crystalline environments is achieved by considering the maximum variations of the torsion angles of the TOT inner ring [22]. An extended comparison including four different space groups is shown in Table 2. In the main, the mean of a given triplet of equivalent torsional parameters does not depend on the crystalline system in spite of widely differing range values. Consequently it has been suggested that these mean values define the stereochemistry of the ideal C_3 TOT molecule [15]. These characteristic torsion angles (see Notes in Table 2) must not be mistaken for parameters necessarily associated with a lower energy TOT conformation. The question has been raised as

Table 2. Range of values (deg.) observed for three equivalent torsion angles in the 12-membered central ring of TOT.

	$P\bar{1}$				$P3_121$		$P6_1$	$Pna2_1$
	I	II	III	IV	V	VI	VII	VIII
a	3	11	2	3	8	10	13	10
b	2	7	1	5	7	10	10	10
c	6	15	7	11	22	25	20	15
d	3	21	5	4	12	12	15	32
e	5	14	10	12	23	28	25	15
f	2	17	3	7	4	8	10	10

The torsion angles are labelled as follows [11] (see Fig. 1 for their equivalents; their "ideal" value referred to in the text is given in brackets): a = O1-C2-C1-C33 [0°]; b = C2-O1-C11-O11 [0°]; c = O1-C11-C12-C13 [60°]; d = C1-C2-O1-C11 [90°]; e = C13-C12-C11-O11 [120°]; f = C2-O1-C11-C12 [180°].
TOT molecules are labelled as follows: I and II in benzene clathrate; III and IV in *trans*-stilbene clathrate; V in (dl)-2-bromobutane clathrate; VI in pyridine and ethanol clathrates; VII in cetyl alcohol clathrate; VIII for unsolvated TOT.

to whether the observed general departure from ideal symmetry is representative of a tendency of part of the molecule to flip over during the process of interconversion to produce the propeller form [15].

Most of the conformations deviate markedly from the ideal C_3 symmetry. As might be expected, the torsion (c and e in Table 2) of the bonds linking the ester grouping to the rigid aromatic moiety exhibit the largest variations. As a whole we note particularly important distortions for unsolvated TOT (VIII) and its conformation (VII) in the cetyl alcohol channel-type clathrate.

On the contrary the least deviations from C_3 symmetry of all the TOT molecules observed thus far are displayed by one (I) of the two independent host molecules in the clathrate with benzene [11], whereas the other molecule (II) is seen to suffer fairly large distorsions. The TOT/benzene clathrate has been visualized as a distorted non-closest hexagonal packing of "spherical" TOT molecules and it is noteworthy that molecule I fits less tightly than II into the lattice space. Consequently the decrease of the environmental packing constraints allows a concomitant closer fit of molecule I with C_3 symmetry.

2.1.4 Shape of the Cavity and Guest Orientation

Space Group $P3_121$:

The cage-clathrate cavity is comprised of eight TOT molecules related in a pairwise manner about a crystallographic twofold axis that typifies the dissymmetry of the cage (Fig. 3). The cages special location limits their number to three in each unit cell with their centers separated by an average distance of 13 Å. In a somewhat idealized manner, the overall van der Waals contour of the cage is that of a deformed ellipsoid with its small axis parallel to *c*, its middle axis

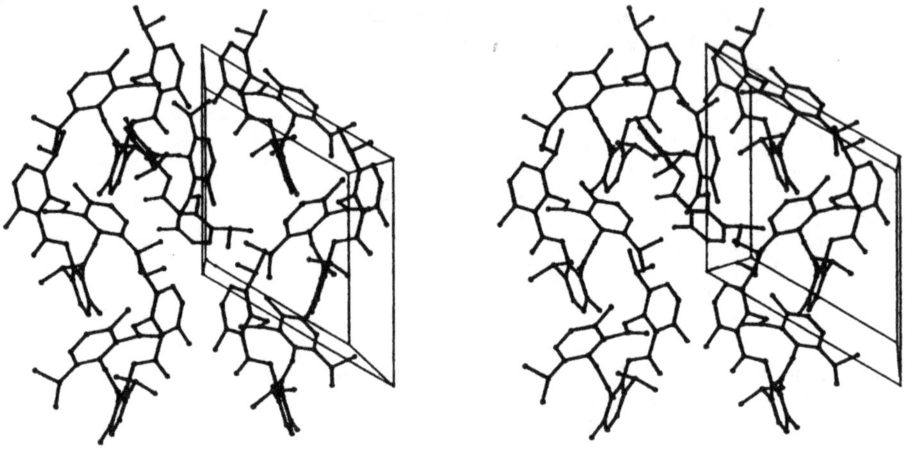

Fig. 3. Stereoview (down the *c* axis) of the packing of the TOT molecules bordering the cavity of a cage-type clathrate (space group $P3_121$). For clarity the top host molecule has been removed. A unit cell is outlined to indicate the crystallographic axes directions. A crystallographic twofold axis lies parallel with the *a* axis seen from down to up

coaxial with a crystallographic twofold axis and its long axis parallel to the *ab*-plane [9a)]. In a typical result with (R,R)-*trans*-2,3-dimethyloxirane as guest molecule the cage axial lengths were estimated at 5.2, 6.1 and 7.0 Å respectively. It has been suggested that these differences prevent random orientation of the guest. The regions of the cage at both ends of the twofold axis intercept differ from one another in shape and nature. At one end a cluster of four isopropyl groups contribute to the formation of a bulge (e.g. near the origin, Fig. 3) which, acting as a local receptor for bulky guest substituents, might have a specific directive effect on the orientation of the enclosed molecule as discussed later. On the opposite side two crystallographic equivalent carbonyl O atoms point into the cage and may be in contact with the guest molecule. It is noteworthy that these O atoms display systematically larger temperature factors than those of the chemically equivalent oxygens in the TOT framework. This feature may be ascribable to a greater freedom of movement in the cage void and has been accounted for by a two-position atomic disorder [9a)].

According to a method described by Lee and Richards [24)], volumes accessible to guest equal to 111, 102 and 96 $Å^3$ have been calculated for clathrates with 2-bromobutane, *trans*-2,3-dimethylthiirane and *trans*-2,3-dimethyloxirane, respectively [9a)]. Comparison with the van der Waals volume of the enclosed guest shows a definite correlation, exemplified by a "constant" volume ratio 0.94–0.97, suggesting a rather close fit of the guest with the receptor. These numerical results depend closely on the "choice" of the van des Waals radii, therefore care must be exercised in interpreting them quantitatively. It seems that the degree of flexibility of the cage to accomodate guests of different sizes is restricted by the small displacements allowed to TOT within the clathrate lattice. Translation, rotation and conformational distortion contribute to the coordinate variations (mean value 0.14 Å, after minimization of the differences in the corresponding TOT positional parameters of two structures [9a]). The relative difficulty encountered in growing crystals of TOT/bromocyclohexane [12] might indicate that the bulky guest (van der Waals volume 121.1 Å^3 [25])) is close to the limit of size of molecules that can be accommodated in cage-type clathrates. The loosened packing of the host molecules reflected by the large-sized unit cell ($a = b = 13.794(1)$, $c = 30.876(1)$ Å) does not decrease the stability of the clathrate crystal but rather affects the growth process in a way that allows a sizeable amount of unsolvated TOT crystals to be formed simultaneously.

X-ray structures (Table 1) demonstrate the involvement of the cage bulge in guest orientation: for all the guests, an heteroatomic component is locked in that particular area nearby the crystallographic twofold axis. This is verified for branched-chain structures as well as for heterocyclic molecules such as *trans*-2,3-dimethylthiirane, pyridine, etc. The orientation of 2-bromobutane within the dissymmetric cage delineated by electron density contours is best visualized in the stereoview, Fig. 4. The particular location of the halogen atom stands out clearly with the C—Br bond extending along the twofold axis direction. It seems that the size of the molecular frame linked to the heterosubstituent is of secondary importance as shown by chlorocyclohexane, where the C—Cl bond adopts an orientation close to that of C—Br in 2-bromobutane. Very interestingly, OH-bearing guests do not get involved in H-bonding with the cage TOT oxygens, as has been observed for 2-butanol and acetic acid [22].

Space Group P $\bar{1}$:

Two crystal structure, with enclathrated benzene and *trans*-stilbene respectively (Table 1), give precise details about the nature of the cavities accessible to guests. These isostructural clathrates are characterized by independent channels

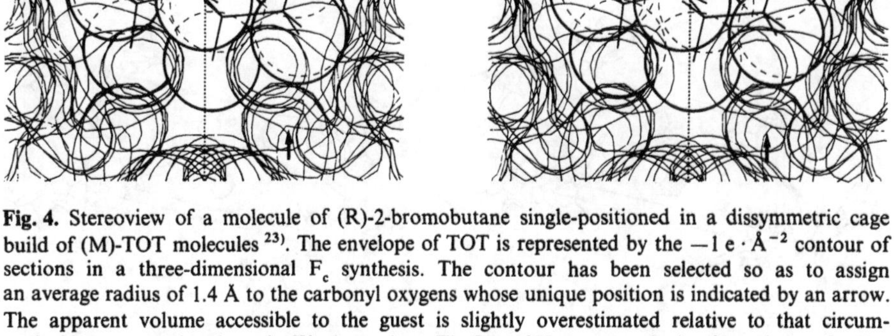

Fig. 4. Stereoview of a molecule of (R)-2-bromobutane single-positioned in a dissymmetric cage build of (M)-TOT molecules [23]. The envelope of TOT is represented by the $-1\ e \cdot Å^{-2}$ contour of sections in a three-dimensional F_c synthesis. The contour has been selected so as to assign an average radius of 1.4 Å to the carbonyl oxygens whose unique position is indicated by an arrow. The apparent volume accessible to the guest is slightly overestimated relative to that circumscribed by a conventional van der Waals envelope. The Br and H atoms are depicted by spheres of radius 2.0 and 1.2 Å respectively. The orientations of the crystallographic axes are as shown in Fig. 3. The dotted line denotes the crystallographic twofold axis

running parallel to the a and b axes. By assuming an average spherical shape of diameter ca 10 Å for the TOT molecules the TOT/benzene clathrate structure has been visualized as a distorted non-closest hexagonal packing of spheres characterized by a periodic stacking of two non-equivalent layers [11]. Indeed, space group $P\bar{1}$ requires the presence of two independent TOT molecules (see Table 2), each of them being surrounded by eight nearest neighbors. Stereoviews of the three independent channels in the TOT/benzene (2:2.5) clathrate (TOTBZ) are presented in Figs. 5 and 6 together with the enclosed benzene molecules. Two channels, T1 and T3, are parallel to a and one, T2, is parallel to b. T1 and T3 exhibit fairly uniform cross-sections along the a axis direction whereas T2 presents a succession of bulges and constrictions. T3 retains the benzene molecules in

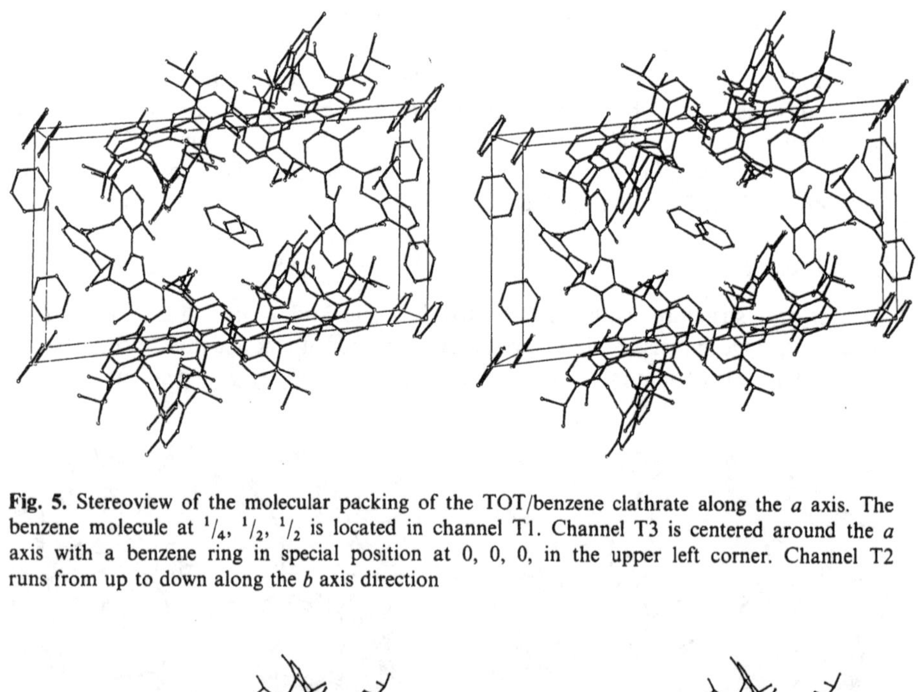

Fig. 5. Stereoview of the molecular packing of the TOT/benzene clathrate along the a axis. The benzene molecule at $^1/_4$, $^1/_2$, $^1/_2$ is located in channel T1. Channel T3 is centered around the a axis with a benzene ring in special position at 0, 0, 0, in the upper left corner. Channel T2 runs from up to down along the b axis direction

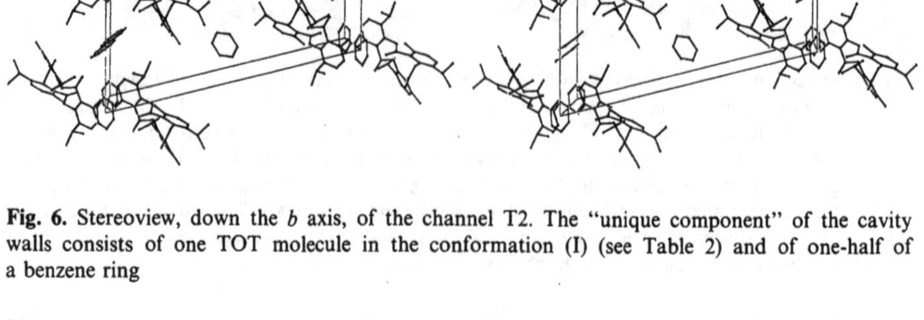

Fig. 6. Stereoview, down the b axis, of the channel T2. The "unique component" of the cavity walls consists of one TOT molecule in the conformation (I) (see Table 2) and of one-half of a benzene ring

special position on symmetry centers where they are in contact with the benzenes locked in T2, close to the point of intersection of the two channels at $^{1}/_{2}$, 0, 0. In this way two independent pairs of benzene rings contribute mutually to the envelopes of the interconnected channels (Fig. 5). This structural arrangement is of significance when compared to the crystal packing of the isomorphous TOT/*trans*-stilbene (2:1) clathrate (TOTSB) where, on the contrary, channel T3 is absent. The removal of the stacks of benzene rings enclosed in T3 must be compensated by a packing rearrangement of the host around the shortest *a* axis. This is evidenced by a contraction of the *b* and *c* axes of TOTSB relative to those of TOTBZ. These structural modifications are undoubtedly facilitated by the flexibility of the TOT framework as suggested by a conformational comparison

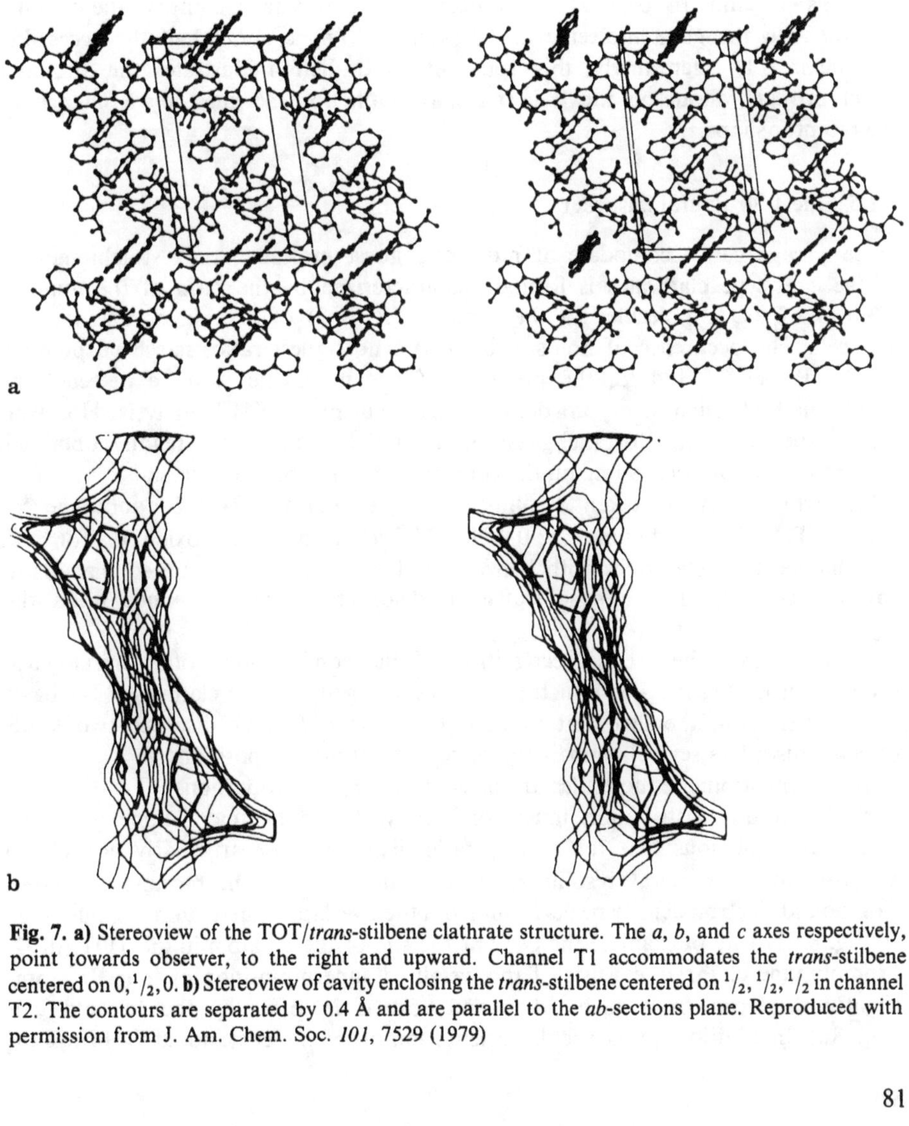

Fig. 7. a) Stereoview of the TOT/*trans*-stilbene clathrate structure. The *a*, *b*, and *c* axes respectively, point towards observer, to the right and upward. Channel T1 accommodates the *trans*-stilbene centered on 0, $^{1}/_{2}$, 0. **b)** Stereoview of cavity enclosing the *trans*-stilbene centered on $^{1}/_{2}$, $^{1}/_{2}$, $^{1}/_{2}$ in channel T2. The contours are separated by 0.4 Å and are parallel to the *ab*-sections plane. Reproduced with permission from J. Am. Chem. Soc. *101*, 7529 (1979)

of I and II with III and IV in Table 2. The packing and orientation of the guests in both clathrates deserve comment. In channel T1 of TOTBZ (Fig. 5) the benzene rings are stacked on top of one another at an angle with the *a* axis direction. The molecules are related by inversion centers and are alternately in van der Waals contact and separated by H ··· H′ distances of 3.15 Å. No disordered guest orientation result from the intervening voids. In the equivalent channel T1 of TOTSB, successive *trans*-stilbene molecules are not coplanar and slightly overlap. They appear in special position on symmetry centers so that the disposition of the phenyl groups is similar to that of the benzene rings in TOTBZ. A similar situation is prevalent in the isomorphous channels T2 of both clathrates, where the guest molecules are more evently spaced. A view of the cavity enclosing *trans*-stilbene in the channel parallel to the *a* axis (T1) is shown in Fig. 7a. The volume left void by the TOT matrix has been calculated and found equal to 570 and 826 Å^3 for channels T1 and T2 of TOTSB, respectively [9b]. In sharp contrast the volume *accessible* to the guest molecule in T2 (see Fig. 7b) is about half the accessible volume in T1. Interestingly, this apparent contradiction illustrates the dramatic effect brought about on the free space available for the guest by a change in the receptor shape.

2.1.5 Guest Structural Disorder

Since a cage can accomodate at most one guest molecule, the "stoichiometry" of the cage-type clathrates is formally characterized by the ratio of the number ($Z = 6$) of TOT molecules to the number ($Z' = 3$) of accessible cages in the unit cell. The occurence of non-stoichiometric host-guest ratios has been pointed out by Powell [26]. Indeed, the presence of vacant cavities can be detected by simple methods such as crystal density measurement or NMR analysis. However precise specific values for the guest occupancy factor are not readily obtained owing to variations depending on the conditions of growth of the crystals [13b]. A few values which may be taken as illustrative are: 0.5 for TOT/2-iodobutane [9a]; 0.74 for TOT/2-bromobutane [13b]; 0.96 for TOT/ethyl-methyl-sulfoxide; 0.8 for both the channel type clathrates with benzene and *cis*-stilbene. It is hinted that a low cavity occupancy may impose statistical disorders of various kinds upon the crystal lattice.

In most cases the precise recognition of the conformation of a chiral guest presents difficulties in X-ray structure analysis of cage-type TOT clathrates. The basic reason resides in the fact that in addition to any *dynamic* disorder two kinds of *static* disorders severely impair the resolution of atomic positions:

a) two enantiomeric molecules are present in varying proportions;

b) the atomic distribution must conform with the twofold symmetry of the cavity. It is obvious that the above static disorders are particularly adverse to a satisfactory structural resolution in crystals grown from racemic solutions. Compound electron density peaks result from the overlap of guest atoms as illustrated in Fig. 8. By way of example the peak at the site of the Br atom in the TOT/(dl)-2-bromobutane clathrate consists of the weighted superimposition of four Br atoms slightly displaced from each other and related in pairs by the twofold axis (Fig. 8a). In addition, the particular orientation of the molecule about C_2 brings the

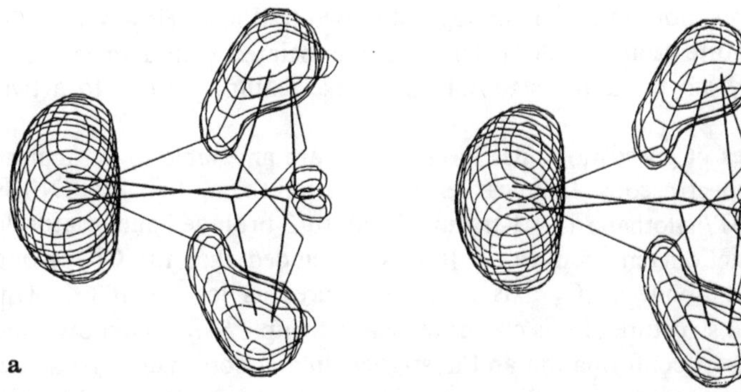

a

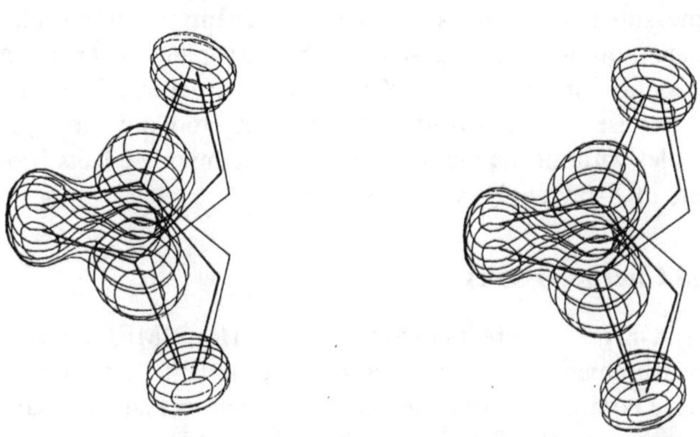

b

Fig. 8. Stereoview of electron density distributions (90% of the total population) calculated from ΔF syntheses in a cage framework consisting of (M)-TOT molecules. a) 2-Bromobutane [13b]. Heavy line: observed preferred (R)-configuration of the X-ray model. Thin line: calculated position of the (S)-enantiomer. b) Ethyl methyl sulfoxide [13c]. The composite density results from the space-averaged contribution of (S) and (R)-enantiomers (thin and heavy lines, respectively) in the approximate ratio 1:11

methyl group in near coincidence with the methyl component of the ethyl group and a single peak results from their overlapping.

The inadequacy of the X-ray model for a straightforward evaluation of the relative contribution of the dynamic and static disorders has been emphasized in the structure analysis of TOT/(dl)-2-bromobutane. Suprisingly, the *R*-ratio test [21)] favored space group $P3_1$ as opposed to space group $P3_121$, in spite of the presence of pairs of TOT molecules shown to be identical within the limits of experimental error and symmetrically related by C_2 axes. In this context it was demonstrated how static disorder could be mimicked by a magnified thermal motion of the Br atom. The Hamilton test gave opposite results with the TOT/(R)-2-butanol clathrate.

It is assumed that coincidence of guest symmetry with cage symmetry should be attained for those guests containing twofold symmetry; therefore optically pure (RR)-*trans*-2,3-dimethylthiirane and (RR)-*trans*-2,3-dimethyloxirane were expected to occupy a special position on the crystallographic two-fold axis. However a high value of the temperature factor of the heteroatoms associated with a smearing of their electron density led the authors to consider these features as due to a two-position disorder [9a)].

Dynamic properties of guest molecules in clathrates are amenable to an analysis by solid state NMR spectroscopy. In an earlier investigation the motional ^{19}F NMR resonance of the TOT/halothane (1,1,1-trifluro-2-chloro-2-bromoethane) clathrate was studied as a function temperature [27)]. It was concluded that the CF_3 group exhibits free rotation about its C_3 axis at temperatures as low as 108 K. Unfortunately the crystal structure of this clathrate (space group *Pbca*) is unknown and assumptions about guest conformation and host-guest interactions had to be made. In a recent paper Ripmeester and Burlinson have studied, by solid state ^{13}C NMR, four cage-type clathrates of TOT formed with *sec*-butyl derivatives as guest molecules [28)]. Their measurements disclose a possible orientational mobility of enantiomeric guests much more important than could be anticipated on examination of the host-guest short contacts in the relevant X-ray crystal structures. These results are pointed out here because they illustrate a method of studying the dynamic disorder of guest molecules without the inconvenience of the mixing effects brought about by static disorders in X-ray analysis.

2.2 Physico-chemical Characteristics

Other techniques than X-ray structure analysis, such as IR, NMR, NQR spectroscopy, enthalpies measurements, etc. can be applied to advantage to investigate the properties of clathrate compounds since in most cases microcrystalline samples are suitable for the measurements. To date very little use has been made of this wide range of methods for investigating TOT clathrates.

2.2.1 NMR Properties

As mentioned previously, conformational changes of TOT in solution have demonstrated the considerable flexibility of the molecule. This result, based upon temperature-dependent NMR signals, has further confirmed the fairly rapid interconversion between the P- and M-propeller configurations *via* helical conformations [29)], over a low racemization barrier of about 21 kcal mol^{-1}. It stands to reason that NMR measurements in solution fail to get information on TOT host-guest interactions. With the advent of solid-state ^{13}C NMR the record of high resolution spectra is now feasible with solid samples using techniques such as magic-angle spinning [30)] and cross-polarization [31)]. This approach lends itself particularly well to the structural study of host-guest compounds since both chemical (intramolecular) and crystallographic (intermolecular) inequivalences are reflected in the nuclear shielding. By way of example, about 20 β-quinol clathrates have been thoroughly examined by ^{13}C NMR spectroscopy [32)]. More recently it was shown, on a more modest scale, that for a series of TOT clathrates of *sec*-butyl derivatives ^{13}C NMR can provide direct indication of chiral recognition, as well as insight into the relative reorientational motion of guest enantiomers [28)]. The latter results were evoked in Section 2.1.5; solid-state NMR and enantioselectivity will be discussed in Section 3.1.5.

2.2.2 Mass Spectral Characteristics of TOT

The mode of mass spectral fragmentation of TOT has been investigated in connection with a study on the use of clathrate compounds for analytical separation [8]. It appeared that the nature of the guest molecule had no effect on the mass spectrum of TOT at the inlet temperature of 180° that corresponds approximately to the melting point of most of the TOT clathrates.

The mass spectral characteristics were as follows: 528 (18, $M^+C_{33}H_{36}O_6$), 352 (85, $M^+(C_{22}H_{24}O_4)$, 324 [33], 176 (67), 161 (33), 148 (100), 133 (7), 120 (5), 105 (14), 77(14) (metastable peaks at m/e 234, 89, 83). The cleavage of two ester bonds (CO—O) of the molecular ion of TOT at m/e 528 results in the

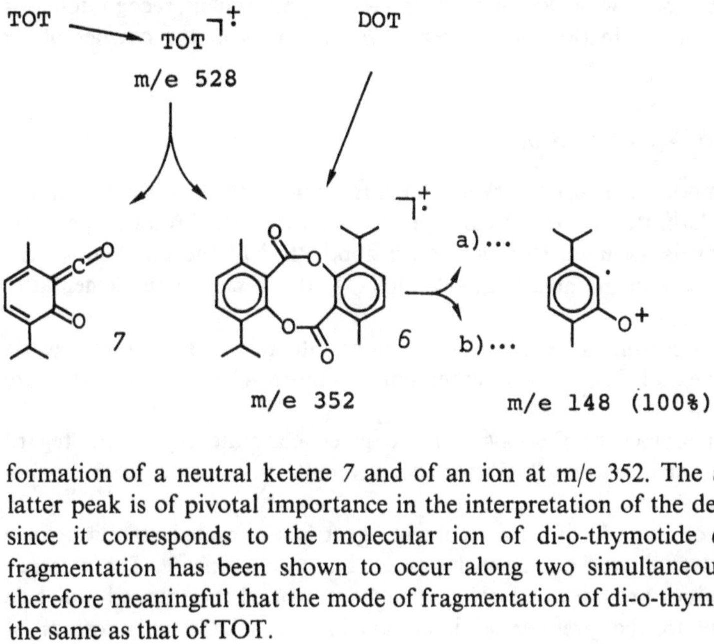

formation of a neutral ketene *7* and of an ion at m/e 352. The assignment of the latter peak is of pivotal importance in the interpretation of the degradation process since it corresponds to the molecular ion of di-o-thymotide *6*; its subsequent fragmentation has been shown to occur along two simultaneous pathways. It is therefore meaningful that the mode of fragmentation of di-o-thymotide is essentially the same as that of TOT.

3 Host-Guest Interaction

The problem of host-guest stereospecific interaction is that of the elucidation of the physico-chemical events that allow the mutual *recognition* of two molecular species leading to a *specific* association. This type of association is of prime importance in biology and pharmacology where it is more conventionally expressed in terms of receptor-substrate complementarity. The nature of the forces that maintain the active conformation of the macromolecular receptor and are involved in the binding of the substrate are fairly well known qualitatively: van der Waals repulsion and dispersion forces, Coulombic forces, interactions with solvatation schells in various environments, hydrogen bonding. However the balanced contribution of these forces to the overall complex formation appears extremely intricate in spite of the fact that several enzymes had their complete three-dimentional structure established by X-ray diffraction to serve as a model.

On several occasions authors have hinted at the analogy between inclusion compounds and biological receptor-substrate complexes. It is beyond the scope of the present review to discuss the analogies (operational and structural) between synthetic macrocyclic receptors [34] exhibiting selectivity in the liquid state and natural enzymatic chemistry. In the opinion of the reviewer any resemblance between crystalline inclusion compounds and biological systems should be appraised with the utmost care and evaluated within the boundaries of their functional behavior. Indeed, clathrate cavities may be poor *stereochemical* models for biological receptors. The crucial difference lies in the fact that biological receptors are assumed to be *flexible* [35] (induced-fit theory) in contrast to the "*rigid*" clathrate cavities (see Sect. 2.1.4) which bestow a lock-key character to molecular recognition. An important additional distinction is the *lack of binding sites* in the cavities of the TOT clathrates.

3.1 Guest Chiral Discrimination

The practically important feature surveyed in this section is the concept of application of TOT clathrates to the separation of enantiomers. Broadly speaking, clathrate formation is assumed to rely upon a good "fit" of the guest molecular shape to the voids within the host lattice. With TOT, this type of complementarity offers two advantages:

a) only non-bonded functional interactions come into play, consequently guests with different shape and size, but with other similar physico-chemical properties are potentially separable;

b) the quasi invariance of the cavity (for a given clathrate type) with regard to varying guest structures may facilitate the rationalization of the phenomenon of stereospecificity.

The possibility of using TOT as a resolving agent has attracted attention since the original report of Powell on the inclusion of 2-bromobutane [36]. The presence of asymmetric cavities provides a chiral environment around the trapped molecule and can give rise to the preferential inclusion of one of the enantiomers of a racemic mixture. A measure of the cavity enantioselectivity is given by the *enantiomeric excess* (e.e.) of the guest in a single TOT crystal of given handedness. Enantiomeric excess and chirality have been determined by various methods: polarometric measurements of the residual optical activity of clathrates solutions after TOT racemization; VPC analysis on chiral phases [37]; NMR using chiral shift reagents. An estimation of the e.e. of caged ethyl methyl sulfoxide has also been carried out by X-ray crystallographic means only [13c]. Figures ranging from 2 to 83% have been observed for the e.e. in cage-type clathrates depending on the nature of the included molecule. On the contrary, the channel-type clathrates have been shown to display invariably low, but significant e.e. values of about 5%.

3.1.1 Host-Guest Correlation of Configuration

Arad-Yellin et al. [9a, 38] have pointed out that a striking correlation exists between the absolute configuration of TOT in a clathrate single crystal and the configuration of the preferentially includes guest enantiomer (Table 3). Thus it was suggested that the assignment of the guest absolute configuration could be possible, within series of comparable guests, on the basis of the sign of the TOT optical rotation measured upon dissolution. Provided the e.e. is significantly ascertained, in particular for the critical low e.e. values, the method could be used to advantage when the formation of diastereomeric derivatives is difficult or impossible for lack of appropriate functional groups. In practice, the large optical rotation of TOT, $[\alpha]_D$ 70°, allows the sign determination on very small clathrate samples of about 1—2 mg.

Conceptually, correlation of configurations occur when the configuration of the preferred enantiomer is such that, in stereochemically related guests, corresponding atoms or groups of atoms are accommodated in the same location of a given receptor framework. The schematic structures presented in Fig. 9 depict the absolute configurations and experimental dispositions of those guests preferentially accommodated by cages built of (P)-(+)-TOT molecules. The features of analogous spatial arrangement of corresponding atoms and similar molecular sizes may be combined in the four molecules: 2-chloro-, 2-bromobutane, ethyl methyl sulfoxide and methyl methanesulfinate. On this basis the (S)-enantiomers of the halobutanes and of ethyl methyl sulfoxide, as well as the (R)-enantiomer of the sulfinate are expected to be preferentially accomodated by (P)-TOT (see *8a*, Fig. 9) as was experimentally verified. These remarks based upon a few of the empirical observations reported in Table 3 assume particular importance in assessing the possibility of anticipating the absolute configuration of the major enantiomer of a structurally simple guest enclathrated in TOT crystals of known handedness.

It has been argued from an interesting experiment that correlation of host-guest configuration should be an intrinsic property of the clathrate and should remain unchanged in every single crystal [9a)]. In this experiment, it was shown that single crystals of the favored association (P)-TOT/(SS)-*trans*-2,3-dimethylthiirane *failed to grow* in a saturated solution of TOT in optically pure (RR)-*trans*-2,3-dimethyl oxirane (the "less favored" component in the (P)-TOT complex grown from a racemic solution). The seed might have induced the incorporation of (P)-TOT together with the available guest enantiomer. Instead, a microcrystalline powder of the (M)-TOT/(RR)-*trans*-2,3-dimethyloxirane clathrate precipitated, though single crystal seeds of TOT/(dl)-*trans*-2,3-dimethylthiirane grew well in solutions of TOT in racemic *trans*-2,3-dimethyloxirane. It has been further suggested, on the basis of the preceding observations, that both e.e. and correlation of configuration are controlled by thermodynamic rather than kinetic factors.

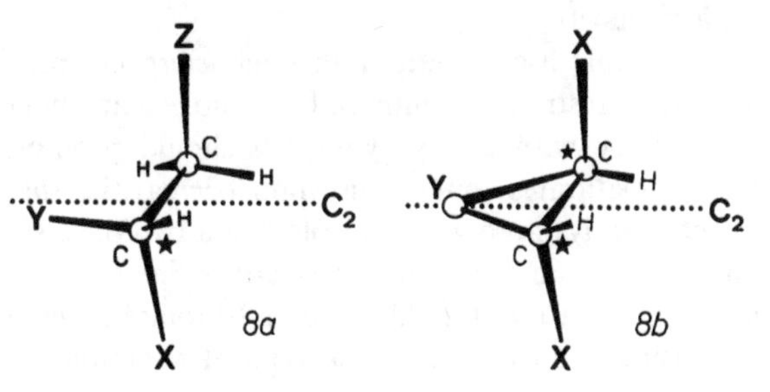

Fig. 9. Configuration and orientation of guests preferentially accommodated by cages built of (P)-(+)-TOT molecules. The dotted line denotes the crystallographic twofold axis.

Table 3. Guest enantiomeric excess and host-guest correlation of configuration in (P)-(+)-TOT clathrates.

Guest	Guest e.e (%)[a]	Guest configuration	Ref.
Cage-type			
2-chlorobutane	32,45	(S)-(+)	38, 39)
2-bromobutane	34,35	(S)-(+)	38, 39)
2-iodobutane	<1	—	9a)
ethyl methyl sulfoxide	83	(S)-(+)	13c)
methyl methanesulfinate	14	(R)-(+)	38)
2-butanol	<5	(S)-(+)[b]	22, 39)
2-aminobutane	<2	—	39)
1,2-dibromopropane	<2	—	23)
trans-2,3-dimethyloxirane	47	(SS)-(−)	38)
trans-2.3-dimethylthiirane	30	(SS)-(−)	38)
trans-2,4-dimethyloxetane	38	—	9a)
trans-2,4-dimethylthietane	9	—	9a)
2,3,3,-trimethyloxaziridine	7	—	9a)
propylene oxide	5	(R)-(+)	38)
2-methyltetrahydrofuran	2	(S)-(+)	38)
Channel-type			
2-chlorooctane	4	(S)-(+)	38)
2-bromooctane	4	(S)-(+)	38)
3-bromooctane	4	(S)-(+)	9a)
2-bromononane	5	(S)-(+)	38)
2-bromododecane	5	(S)-(+)	38)

[a] Single crystals grown from racemic solutions of guests.
[b] Grown from optically pure guest (see also Ref. 20).

3.1.2. Stereochemical Approach

In view of the earlier work of Lawton and Powell [26)] on unit cell dimensions and contents of channel-type clathrates, the guest molecules are expected to be severely disordered within the "cylindrical" cavities. First, they may be packed "head-to-head" or "head-to-tail"; second, the length of the chain-like molecules is unlikely to match the unit translation imposed by the ordered host lattice. As a consequence, host-guest functional interactions will differ within successive unit cells and a dramatic decrease of chiral discrimination is bound to ensue for lack of unique stereospecific complementarity.

The qualitative interpretation of chiral discrimination through stereochemical data rests on the premise that, within restrictions imposed by molecular shape and size, coincidence of guest symmetry with cavity symmetry should promote higher e.e. values. Since the cage-type clathrates contain dissymmetric cavities they are expected to discriminate more efficiently those guests displaying a twofold axis. This is indeed the case for the thiirane and oxirane derivatives in Table 3. On the other hand, the high enantiomeric purity of 2-chloro- and 2-bromobutane is rather unexpected on symmetry consideration alone. In the type of enantiomeric

Table 4. Close contacts[a] (Å) of enantiomers with TOT host molecules

	2-Bromobutane[b]		*trans*-2,3-Dimethylthiirane	
$\langle\Delta\rangle$	0,058(15)	0,093(17)[c]	0,099(17)[d]	0,116(32)[e]
n	9	8	17	12

[a] The contacts are characterized by the difference Δ = (sum v.d.W. radii) — (host ··· guest interatomic distance); $\langle\Delta\rangle$ is the average value taken over n contacts with $\Delta \geqq 0$ [40]. [b] Contacts minimized by optimization of guest orientation, position and internal degree of freedom. [c] After Br/CH_3 exchange. [d], [e] X-ray crystallographic values for the major and minor enantiomer, respectively.

disorder described precedingly (Sect. 2.1.5) on the basis of X-ray crystallographic data the enantiomers of 2-bromobutane are approximately symmetry-related by reflection in a mirror whose orientation leaves their terminal atoms fixed in similar locations of the cage. This type of static disorder is best illustrated in Fig. 8a. There is however another possible form of static disorder, termed "ligand-exchange" disorder [9a)], which has been discussed in the particular case of 2-bromobutane. In this approach the guest molecule is inverted by the exchange of Br and CH_3, the other atoms bonded to the asymmetric carbon remaining fixed. Thus, in this model both Br and one terminal CH_3 group would have a different environment for each enantiomer. When van der Waals host ··· guest contacts were minimized for both enantiomers in their respective environment after ligand-exchange, comparable values were obtained (mean values are reported in Table 4). The actual contacts of both enantiomers of *trans*-2,3-dimethylthiirane with the cage are also comparable. In this complex the S atom of both enantiomers remains fixed close to the cage twofold axis (see Fig. 9). The actual occurence of ligand-exchange disorder seems to have been disproved by careful X-ray structural analysis.

The orientation of the preferred enantiomer of 2-bromobutane places Br in a hydrophobic environment characterized by short Br ··· H contacts, whereas the CH_3 groups are locked in the vicinity of the TOT carbonyl oxygens. Through ligand-exchange, one O atom is brought in contact with Br and a dominant Coulombic repulsion arises between the interacting atoms. It seems reasonable to admit on a qualitative basis that this latter type of interaction might have a directive effect on the final equilibrium position observed for all the guest heteromolecules included in the TOT cage-clathrates. However the necessity of taking into account an entire set of non-bonded atom ··· atom potentials to rationalize the packing energy differences between diastereomeric host-guest associations is demonstrated in the following section.

3.1.3 Semi-empirical Study of Enantioselectivity

It results from the approach presented in the preceding section that the qualitative manipulation of "lock-and-key" models does not provide adequate information about the relative energies involved in the diastereomeric host-guest interactions, hence it does not seem possible to predict the preferred guest enantiomer. However the relative complementarity of a pair of enantiomers to the shape of the cage is amenable to a semi-quantitative estimation under the assumption that the enantiomer giving rise to the least repulsion energy with the cage will be preferentially enclosed. The first attempt to explore this problem has been undertaken by Gerdil and Allemand [39] on the basis of the calculation and comparison of the minimum energies for the inclusion of enantiomers within a rigid cage of given chirality. An energy difference was defined as $\Delta RS = |E(S) - E(R)|$, where E(S) and E(R) are the minimum non-bonded potential energies for the inclusion of the (S) and (R) enantiomer, respectively. Most simply, it was believed that chiral recognition would be roughly proportional to the magnitude of ΔRS.

The method was primarily illustrated by the interpretation of the fair chiral recognition of 2-bromobutane (e.e. 35%) as opposed to the unselective enclathration of 2-butanol (e.e. $<5\%$). The position of the guest molecules was optimized by the use of the PCK6 program [41] which allows the minimization of the packing energy of several distinct rigid bodies in a crystal lattice. Calculation showed that the diastereomer (M)-TOT/(R)-2-bromobutane is more favorable by 3.5 $kcal \cdot mol^{-1}$ than its congener (M)-TOT/(S)-2-bromobutane, in good qualitative agreement with the experiment. As pictured in Fig. 10, the final calculated position of 2-bromobutane converged on that observed for the crystallographic model. Inspection of the close intermolecular contacts around the methyl groups calls for some rotational freedom that was neglected in the calculations, by which is meant that limited rotations of the methyl do not bring about critical variations of the packing energy. Both 2-bromobutane and 2-butanol are included in very similar environments, however the alcohol possesses an additional degree of freedom, defined by the torsion angle C—C—O—H (Θ). The ease of inclusion of 2-butanol was shown to depend very much on the internal orientation of the O—H bond. The packing energy profiles of the (R) and (S) enantiomers differ markedly but the curves intersect for two values Θ_1 and Θ_2. This is best illustrated by plotting the algebraic difference of the respective potential energy curves as depicted in Fig. 11b. By comparison with the strain energy profile of 2-butanol as a function of Θ (Fig. 11a)

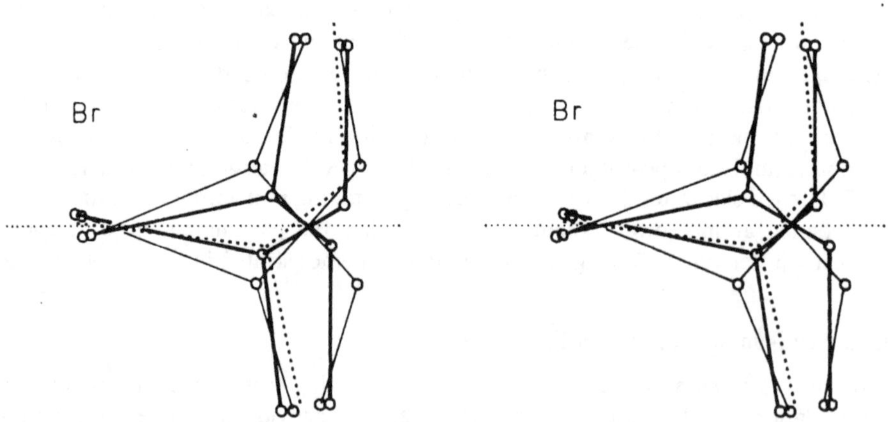

Fig. 10. Stereoview, down the *c* axis, of the calculated positions for (R)- and (S)-2-bromobutane (heavy and thin line, respectively). Dashed line: X-ray model (one equivalent position shown). Dotted line: crystallographic twofold axis

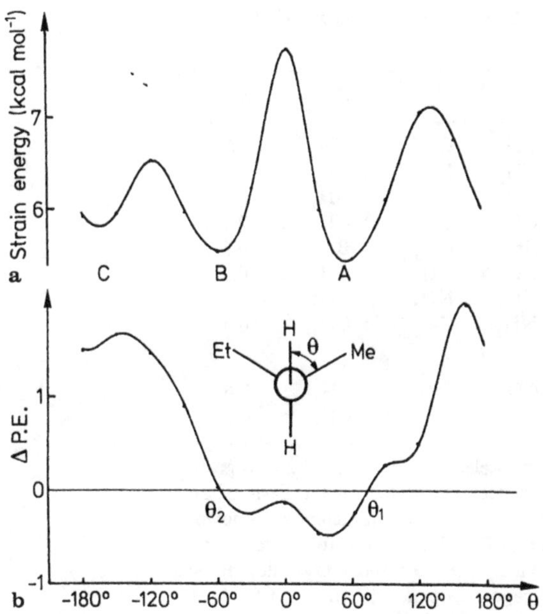

Fig. 11. a) Strain energy of 2-butanol *versus* torsion angle C—C—O—H (Θ). **b)** Inclusion of a rotational isomer of 2-butanol in a cage of the (M)-TOT lattice: [potential energy of (S) *minus* potential energy of (R)] *versus* Θ

it is seen that both values Θ_1 and Θ_2 also correspond to Θ-values pertaining to minimum energy conformations A and B of 2-butanol. The fundamental outcome is that either (R)- or (S)-2-butanol can be enclathrated at the expense of about the same energy without any drastic internal molecular strain, thus pointing to a greatly reduced enantioselectivity of the cage, as observed experimentally.

Predictions for unknown crystal structures are facilitated by the fact that the cage-type clathrates are isostructural. A simple measurement of the cell parameters permits an adequat optimizing of the cage dimensions provided that a rigid average conformation is used for the TOT molecules, which are then allowed to move independently during the optimization procedure. This semi-empirical approach has been further extended to several guest components represented by the frameworks *8* (Fig. 9); detailed results are reported in Table 5 [42]. In the calculations of the conformational energy variations of the guests, allowance was made in a way similar to that used for the constrained 2-butanol model. By inspection of Table 5 it is seen that the energy differences ΔRS can be arranged in two groups of values differing consistently in magnitude:

a) $\Delta RS \geqq 2.7 \text{ kcal} \cdot \text{mol}^{-1}$;

b) $\Delta RS \leqq 0.5\text{–}0.8 \text{ kcal} \cdot \text{mol}^{-1}$.

The guests which are substantially discriminated (with the exception of 2-iodobutane) are found in group (a), whereas those for which low or vanishing small chiral recognition has been observed are contained in group (b). Thus it has been stated that in first approximation the calculated energies reported in Table 5 are accurate enough to reflect the likely occurence of absence of chiral recognition for a given guest molecule [42]. The conformational behavior of the guest molecule is also of great significance in that it has been shown to give stereochemical clues to the impressive variations of

Table 5. Calculated energy difference (ΔRS) for the inclusion of R and S enantiomers in a cage of given handedness.

Guest (8a)	X	Y	Z	ΔRS[a]
2-chlorobutane[b]	: CH_3	: Cl	: CH_3	: 3.2
2-bromobutane	: CH_3	: Br	: CH_3	: 3.5
2-iodobutane[c]	: CH_3	: I	: CH_3	: 2.6[d]
ethyl methyl sulfoxide	: CH_3	: O[g]	: CH_3	: 2.7
1,2-dibromopropane[e]	: Br	: CH_3	: Br	: <0.5
2-butanol	: CH_3	: OH	: CH_3	: <0.5
2-butylamine[c]	: CH_3	: NH_2	: CH_3	: <0.5
2-amino-1-butanol[c,f]	: NH_2	: CH_2OH	: CH_3	: <0.5
Guest (8b)				
trans-2,3-dimethyloxirane	: CH_3	: O	: CH_3	: 2.8
trans-2,3-dimethylthiirane	: CH_3	: S	: CH_3	: 2.3

[a] Absolute value (kcal · mol^{-1}). The enantioselectivity of the clathrates is given by the e.e. values reported in Table 3. [b] ΔRS calculated on the basis of an X-ray crystallographic model [43]. [c] Only the unit cell parameters are known. [d] The discrepancy between the significant ΔRS value and the absence of chiral discrimination might stem from an inadequate theoretical crystallographic model that neglects the observed low cage occupancy (Section 2.1.5). [e] Br positions ascertained from X-ray structure analysis [44]. [f] X and Y may be exchanged with each other. [g] In 8a the methine group is replaced by a S atom.

the enantioselectivity with respect to the included structure. Expressed in qualitative terms, it has been concluded that the presence of a "rigid" rotor alongside the crystallographic C_2 axis (structure *8a*) tends to cause a substantial decrease of the ΔRS values with a concomitant levelling of the chiral discrimation.

3.1.4 Amplification of Chirality on Crystallization

The amplification of chirality from achiral, racemic [45] or prochiral systems is embodied in the fascinating problem of the emergence of chirality during molecular evolution [46]. Amplification of chirality comes whithin the scope of clathrates enantioselectivity as a fundamental mechanism that enhances the enrichment in one encaged enantiomer starting from a slightly unbalanced mixture of chiral components. As illustrated in Fig. 12, the initially low e.e. of the channel-type TOT/2-bromooctane clathrate can be considerably increased when optically active solutions of the guest are used [47]. Thus high enantiomeric purity may be easily attained on repeated crystallizations by reprocessing the enantiomerically enriched guest extracted from the crystals of the preceding crop.

In an attempt to rationalize the phenomenon of chirality amplification it has been assumed [47] that during the first stage of crystal growth the host aggregates more rapidly than guest, leading to the inclusion of the latter molecules into partially formed channels. Under these circumstances the already included molecules may interact favorably with approaching guests of the same chirality, in a way that increases the chiral discrimination beyond that of the chiral host "alone". Thus, as the starting concentration of one enantiomer increases, the probability of homochirality

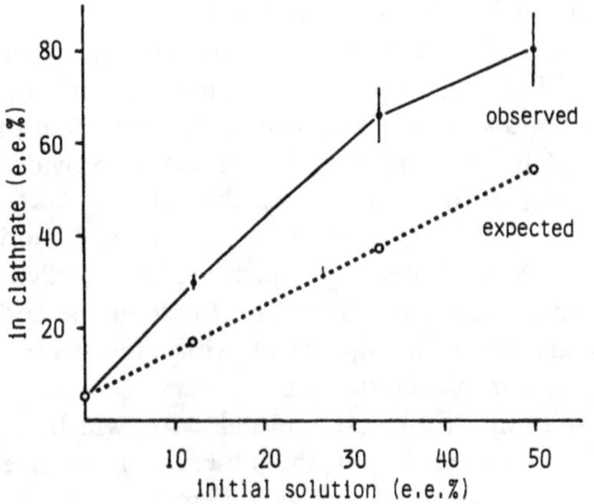

Fig. 12. Amplification of guest optical purity on TOT crystallization from optically active 2-bromooctane [47]. The "expected" e.e. is calculated from the 5% intrinsic chiral discrimination. Error limit ±10% of value

increases. An additional cooperative factor may operate at the solution/crystal interface where homochiral guest ··· guest interactions between the preferred enantiomers may exclude the minor enantiomers far below the concentration factor. A related viewpoint has been discussed by Wynberg concerning the consequences of interactions between enantiomers in asymmetric reactions [48].

The seeding of a TOT solution in the guest to be resolved is another aspect of the amplification of chirality which entails practical applications when a large quantity of optically active guest is needed. A powdered single clathrate crystal is added to a hot TOT solution which is cooled and stirred for a period of time. Fine crystalline powders of the clathrate is thus obtained containing a large enantiomeric excess of crystal of one handedness. When this has been done with 2-bromobutane, the optical purity of the extracted guest was 85% of that obtained in a single crystal [9a].

3.1.5 Chiral Discrimination and Solid-State ^{13}C NMR Spectroscopy

NMR spectra obtained from solids are quite different from those obtained from liquids and yield different information. In particular further insight into the nature of the motional behavior of lattice components can be gained by solid-state NMR spectroscopy.

It has been shown that ^{13}C spectra afford a straightforward indication of chiral discrimination for several TOT clathrates of secondary-butyl-compounds, including quantitative determination of enantiomeric excesses [28]. Also, differences between the dynamic states of the enantiomers are revealed by the MNR measurements. Powdered crystalline samples of the clathrates (conglomerates) were prepared with 2-chloro-, 2-bromo-, 2-iodobutane and 2-butanol as guests. In the solid, the lines of the host molecule become triplets in accordance with the loss of threefold molecular symmetry. The observable guest resonances are as follows (C(1) resonance is obscured by the TOT spectrum at low field): C(2) gives rise to distinct doublets for all the guests, and C(4) only for the halo-derivatives; the remaining lines, including C(3) resonance, are singlets. In consideration of the correlation of configuration described in Section 3.1.1 (Table 3) the major lines of the C(2) and

C(4) guest doublets were assigned to (P)-TOT/(S)-guest and (M)-TOT/(R)-guest associations. The assignement was separately confirmed by the clathrates prepared from optically pure 2-bromobutanes [49] whose spectra are consistent with the absence of a minor enantiomer. Thus, solid-state NMR affords a direct method of estimating enantiomeric excesses in mixed diastereomeric crystalline systems provided that the dependence of the line intensities on the cross-polarization time is taken into account. From the peak size of the C(4) methyl resonance, e.e. of 47% and 35% were found for 2-chloro- and 2-bromobutane, respectively, in excellent agreement with values reported by Gerdil and Allemand [39]. For 2-iodobutane and 2-butanol, the NMR measurements are again in agreement with previous results [9a, 39], indicating almost equal population of enantiomers.

Owing to the occurrence of line splittings for guest molecules for which no chiral recognition is exhibited, it was concluded that the observation of line splittings has no direct bearing on the presence of energy differences between the related diastereomeric "cavity/guest enantiomer" associations in the crystal.

The dipolar dephasing technique [50] was used to advantage to obtain further insight into the dynamic properties of the encaged sec-butyl compounds and into their chiral discrimination as well [28]. The ratio I_{DD}/I of ^{13}C line intensities measured under dipolar dephasing (DD) and normal cross-polarisation/magic angle spinning conditions is a semiquantitative measure of the dynamic state of the molecular group. For locked methylene or methine C atoms $I_{DD}/I \approx 0$, whereas enhanced molecular motion increases the I_{DD}/I ratio, up to a maximum value of 1. For 2-chloro- and 2-bromobutane, C(2) has $I_{DD} \approx 0$ for the major enantiomer and a value of ca. 0.4 for the minor enantiomer. Similarly the ratios for the C(4) atom have values of ca. 0.6 and ca. 0.9 for the major and minor enantiomers, respectively. These results suggest that the preferred enantiomer fits rigidly into the cage, wherease the minor component is submitted to a reorientation process analogous to a single-axis rotation. The C(2) atoms of both enantiomers of 2-butanol have $I_{DD}/I \approx 0.4$ so that they both display the same degree of motional freedom in the cavity. The conclusion arrived at on the basis of the solid-state NMR observations was, therefore, that enantioselectivity seems reflected more by the dynamic guest properties (static major enantiomers, mobile minor enantiomers) than the equilibrium guest positions determined by X-ray diffraction. The NMR model is of great interest because it is somewhat at variance with the X-ray structural models for chiral recognition, as discussed in the preceding sections.

3.2 Other Stereospecific Interactions

Besides the chirality-discriminating property exhibited by the enantiomorphous TOT clathrates, the feature of molecular shape and/or size recognition has been used for the relative separation of hydrocarbons from mixtures in which they occur [51]. The clathrates were formed by crystallization from hydrocarbon blends of known composition. A small crystal of each complex was decomposed by heating and the vaporized component guests analyzed by gas chromatography. A few typical results are reported in Table 6. Little consideration was given to the crystallographic aspects of the various clathrates investigated, consequently the type of cavity occurring in several of the complexes can only be conjectural. It will be noted that chain

Table 6. Composition of hydrocarbon component of TOT complex crystals grown from hydrocarbon blends

	Hydrocarbon blend[a]	Guest % composition of component (wt. %)
a	2,2,3-trimethylbutane	1.3
	3,3-dimethylheptane	1.4
	3-ethylpentane	1.4
	2,3-dimethylpentane	5.0
	2,2-dimethylpentane	5.9
	3-methylhexane	7.4
	2,4-dimethylpentane	14.1
	2-methylhexane	29.6
	n-heptane	33.0
b	2-methyl-2-pentene	24.4
	cis-2-hexene	25.8
	trans-2-hexene	50.0
c	benzene	98
	n-hexane	2
d	cyclopentane	98
	n-pentane	2

[a] The initial blends are composed of equal volumes of hydrocarbons.

or slightly branched chain hydrocarbons are more easily enclathrated than the more highly branched chain hydrocarbons (blend a). Also preference is given to the inclusion of cyclic structures with respect to homologous chain-like frameworks (blends c and d). It was concluded that cage-type clathrates are formed in preference to the channel-type clathrates when the possibility of forming alternative systems arises.

The formation of a TOT inclusion compound with thymol has been reported [52]. The compound was prepared by heating a mixture of TOT with an excess of thymol till complete solution occurred, then the mixture was allowed to solidify on cooling and the thymol content in the inclusion compound analyzed by a colorimetric method. A thymol concentration of 16.1 % was consistently measured in a number of independent samples and cross checked by GLC analysis (m.p. of the inclusion compound 154°). The thymol content is higher than that usually expected (12.4 %) for the host: guest ratio (2:1) found for the cage-clathrates.

3.3 Guest Structural Modification in Cage Clathrates

In the last twenty five years there has been a number of studies devoted to the reactions of the organic solids subjected either to a physical agent (light, heat) or a chemical one which can be a solid, a liquid or a gas [53]. The molecular constituents of a highly ordered medium such as a crystal lattice may undergo physical or chemical modifications which are far more selective than those experienced in solution or in the gas-phase. In the bimolecular crystalline edifices of the clathrates, attention is turned to the guest molecules as incipient reactive species. For example, it has recently been reported that inclusion polymerization of prochiral dienes monomers in the channels of apocholic acid clathrates proceeds under topochemically controlled conditions with a high degree of asymmetric induction [54]. Since the guest reactants are locked in a highly anisotropic rigid environment, high specificity may be expected for various processes. Few possibilities have been exploited along this line in the case of the TOT clathrates.

3.3.1 Chirality Transfer on Photo-Oxygenation

The first case of a bimolecular heterogeneous reaction stereocontrolled by the chirality of a crystalline host receptor has been recently reported [55]. The substrate was a prochiral olefin included in the cavities of a TOT clathrate and subjected to the action of an incoming chemical agent. As suitable prochiral guest molecules (Z)- and (E)-2-methoxybut-2-enes (*9* and *11*) were envisaged. They both react with singlet oxygen to yield allylically rearranged hydroperoxydes (*10* and *12*) in solution. A suitable cage-type clathrate (space group $P3_121$) was isolated with the Z-isomer as guest molecule, whereas the less favorable E-isomer only furnished an achiral triclinic clathrate. The Z-guest adopts a static, twofold disordered arrangement with the crystallographic twofold axis bisecting the olefinic bond and passing through the O atom (Fig. 13a).

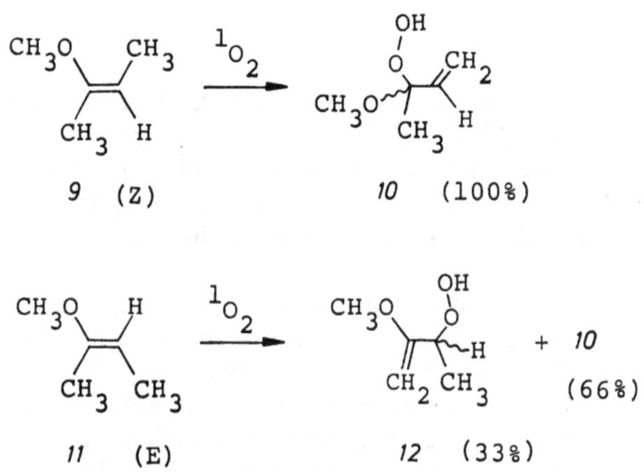

Experimentally, crystal crops of opposite handedness were separated manually and submitted to photo-oxygenation under the same adequate conditions. After racemization of TOT on solution in C_6D_6, polarimetric measurements revealed a residual optical activity indicative of the occurrence of an asymmetric reaction. Although the enantiomeric excesses in the experiments were not determined, the similarity of amplitude and complementarity of sign of the optical rotations were significant as demonstrated by a typical experiment (an e.e. of 100% is assumed in the calculations):

$$\text{(P)-(+)-TOT}/9 \xrightarrow{^1O_2} (+)\text{-}10 \quad [\alpha]_{546}^{20} = +120° \pm 20°$$

$$\text{(M)-(−)-TOT}/9 \xrightarrow{^1O_2} (-)\text{-}10 \quad [\alpha]_{546}^{20} = -146° \pm 20°$$

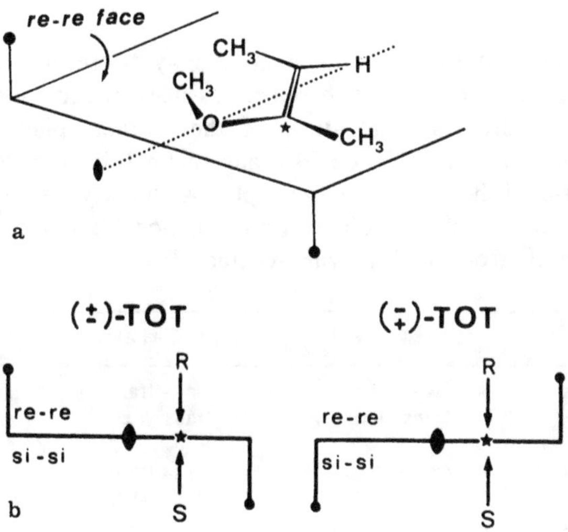

Fig. 13. a) Idealized view of the orientation of the prochiral guest *9* relative to the crystallographic two-fold axis (dotted line). The dissymmetric environment is schematized by the two symmetrically equivalent points linked to the olefinic mean molecular plane. **b)** Diastereomeric relationships involving the creation of an asymmetric carbon at the starred center in *9*. The arrows denote the opposite directions of attack by singlet oxygen at the enantiotopic olefin faces and the configuration (R or S) of the generated hydroperoxyde *10*

It was inferred that the incoming singlet oxygen permeates the host lattice and reacts with the prochiral olefin by two diastereomeric transition states (Fig. 13b), leading to an apparent chirality transfer from the dissymmetric environment of the cage to the hydroperoxyde under formation.

3.3.2 Isomerization

Racemization:

The temperature dependence of a physical phenomenon such as racemization has been investigated with a view to provide clues about the involvement of the cage rigidity in chiral discrimination [9a, 22]. The work was aimed at the determination of the change brought about on the barrier of racemization of an easily inverted guest molecule by enclathration. The relevant experimental features were provided by the racemization of methyl methanesulfinate [$CH_3S(O)OCH_3$] within the cage: no racemization took place in clathrate crystals heated to 115 °C for 12 hours, whereas complete racemization occurs in solution under the same conditions. This notable stabilization towards racemization in the cavity vanishes above a certain temperature and complete racemization resulted on heating the clathrate at 125 °C for 12 hours, with a corresponding levelling of the e.e. from 14% to zero. Powder diagrams before and after the heating ascertained that the cage clathrate structure had not been destroyed, from which it was assumed that enantiomerization had proceeded within the cage of the host lattice. The inversion process at higher temperature was interpreted as due to a loss of chiral discrimination related to the increased thermal motion of the TOT atoms bordering the cage. It was further assumed that, on cooling, the racemization barrier of the guest is passed before the "thermal barrier" of TOT, over which the chiral recognition of the cage is restored.

Photoisomerization:

In olefin derivatives the π component for the double bond may be reversibly broken photochemically. *Cis-trans* interconversion is therefore possible by irradiation *via* an excited transition-state. A circumstantial study of the *cis-trans* photoisomerization of enclathrated stilbene (PhCH=CHPh) and methyl cinnamate ($PhCH=CHCOOCH_3$) has been published [9b]. Although a photostationary state is readily reached in solution, the behavior of both olefins on irradiation is markedly modified in the solid-state as appears from the following scheme [22]:

	solution	guest crystal	TOT-clathrate
stilbene	cis $\rightleftharpoons$ trans 80% 20%	cis $\nrightarrow$ trans trans $\nrightarrow$ cis	cis → trans trans $\nrightarrow$ cis
methyl cinnamate	cis $\leftrightarrows$ trans 50% 50%	cis → trans [56] trans $\nrightarrow$ cis	cis $\leftrightarrows$ trans 50% 50%

It stands out that the photochemistry is also *dependent of the crystalline phases* where the topochemical relations between nearest-neighbor molecules are not the same in the pure guest crystals and in the TOT clathrates. When the TOT/*cis*-stilbene clathrate was irradiated, photoconversion to *trans*-stilbene occurred smoothly with the additional formation of small amounts of phenanthrene and of an unidentified product. On the contrary the TOT/*trans*-stilbene clathrate remained unchanged when irradiated for long periods of time. In the majority of cases, crystalline *cis*-cinnamic acid derivatives are entirely converted to *trans*-isomers on irradiation and the *trans*-to-*cis* isomerization is not observed [57]. However *both* TOT clathrates with the *trans*- and *cis*-isomers of methyl cinnamate photoisomerized under similar conditions. It is noteworthy that these clathrates and those of the stilbenes (Sect. 2.1.4) are isomorphous.

The first stage of most photochemical solid-state reactions does not require an initial loosening of the molecular packing at the reaction site as has been assumed for thermal reactions [58]. Accordingly the photoisomerization *via* TOT enclathration has been envisaged as an essentially "cage-controlled" process consistent with the experimental observations. The facile *cis*-to-*trans* photoconversion of stilbene was reported to be favored by the low cage occupancy in the TOT/*cis*-stilbene clathrate and by the coincidence of the *trans* guest symmetry with that of the cavity. This symmetry matching was regarded as a major element of control over the reaction since the non-centrosymmetric *cis*-stilbene never develops the same complementarity with the centrosymmetric cavity as does the ordered product molecule; hence the marked preference of TOT for *trans*-stilbene rather than for *cis*-stilbene is evident. The stabilizing directive influence of guest-cage symmetry coincidence is further supported by the reaction pattern of the methyl cinnamate isomers which are disordered in their respective cavities and both undergo photoisomerization in the clathrates.

Iodine catalysis can also promote full (thermal) *cis*-to-*trans* conversion of enclathrated *cis*-stilbene. The absence of any sign of partial conversion supports the idea that the overall process is not biased by the presence of particular sites

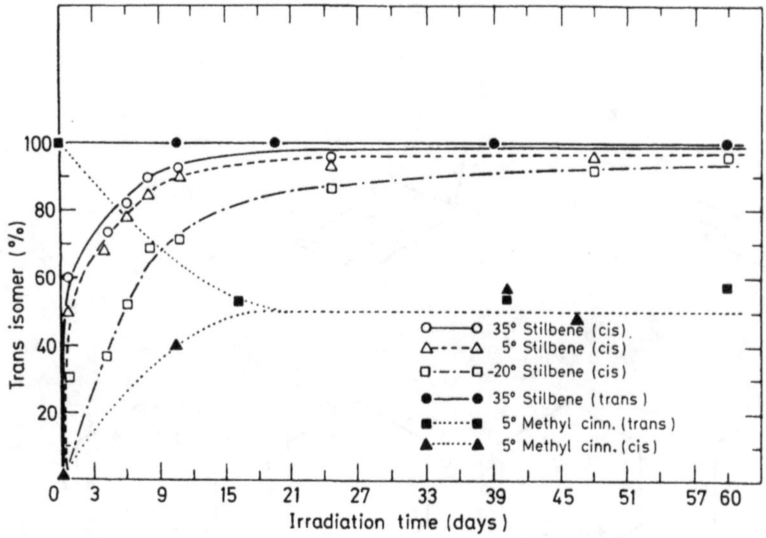

Fig. 14. *Cis-trans* photoisomerization of stilbene and methyl cinnamate in TOT clathrates. Reproduced with permission from J. Am. Chem. Soc. *101*, 7529 (1979)

extraneous to the "normal" cavities, where "exposed" guests would be easily isomerized. Exposure of TOT/*trans*-stilbene to iodine vapor caused no detectable isomerization.

3.3.3 Stabilization of Strained Conformers

The attraction forces between host and guest are necessary to the cohesion of the TOT clathrate lattice since crystal decomposition happens to be the inescapable fate met by the clathrates on desolvation. In a few instances the included guest molecule may assume a strained (unusual) conformation that is beneficial to the *overall stability* of the clathrate lattice. Conversely, the host lattices may be considered as media for stabilizing *unstable guest* species [59].

Several IR and Raman spectroscopy studies of the thiourea inclusion compounds of monohalocyclohexanes have been reported [60]. The predominance of *axial* conformers in the cavities has been demonstrated, in contrast with the preferred *equatorial* conformation in the liquid or gaseous phases. No X-ray structures of the thourea complexes have been reported.

The TOT-clathrates with chlorocyclohexane, bromocyclohexane and 2-chlorotetrahydropyran have been prepared and investigated by X-ray crystallography and by IR spectroscopy [12]. The crystal structure of TOT/chlorocyclohexane at 158 °K (Table 1) disclosed the inclusion of two energetically disfavored conformations of the guest: an *axial*-Cl chair and an *axial*-Cl boat conformation distributed statistically in the ratio 2:1 over the available sites. There was no evidence for the presence of the thermodynamically most stable *equatorial* isomer. The precise crystal structure of the guests was partially impaired by the weighted superimposition of the two molecules and the local static twofold disorder in the cage: improved geometries

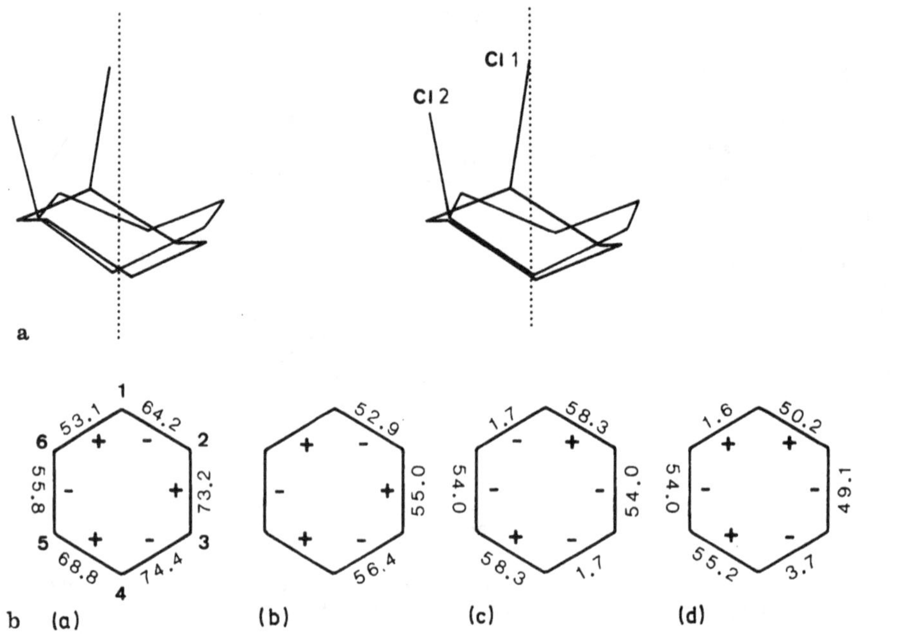

Fig. 15. a) Stereoview, along the *c* axis, of the encaged conformations of chlorocyclohexane. Dotted line: crystallographic twofold axis. For clarity the guest molecules are shown single-positioned. **b)** Endocyclic torsion angles of chlorocyclohexane in various conformations: (a) major observed; (b) major calculated, C_s symmetry; (c) minor observed; (d) minor calculated. The axial Cl atom is linked to atom C(1). Reproduced with permission from J. Incl. Phenom. *3*, 335 (1985)

were calculated using force field methods [61]. A stereoview of the guest conformations and a comparison of their observed and calculated endocyclic torsion angles are given in Fig. 15. Optimization of the internal coordinates of the observed major conformer converged to a minimum energy conformation displaying C_s molecular symmetry (strain energy $E_s = 7.2$ kcal mol^{-1}); the observed chair is slightly distorted. The best fit ($E_s = 14.1$ kcal mol^{-1}) with the observed boat isomer was not reached by optimization of the experimental internal parameters but from those of an "ideal" boat structure with additional axial Cl-substitution. As for acyclic halogenated guests (Sect. 2.1.4) the C—Cl(1) bond axis of the major conformer lies roughly parallel to the twofold axis direction with Cl(1) locked in the "bulge" of the cavity. The Cl(2) atom is displaced from the more favorable location occupied by Cl(1), to the benefit of a better fit of the boat ring atoms with the cavity. Steric repulsion between Cl(2) and the cage is minimized by a local disordered rotation of a TOT isopropyl group. It is noteworthy that the *axial*-boat conformation trapped in the clathrate crystal has never been detected in the liquid-phase.

Infrared Spectra:

The IR spectra of the monohalocyclohexanes is well documented [62]; the assignment of IR absorption bands of pure TOT has not been attempted at the present time.

Table 7. Characteristic IR absorption bands (cm^{-1}) in the region 850-600 cm^{-1} for the pure guest and its TOT clathrate

	eq.[a]	ax.[b]	eq.[b]	ax.[b]
chlorocyclohexane	818	807	732	685
clathrate	—	811	—	684[c]
bromocyclohexane	810	804	687	658
clathrate	—	804[d]	—	660[e]
2-chlorotetrahydropyran	—	813	—	699
clathrate	—	814[f]	—	696

[a] Ring stretching frequency. [b] C-X stretching frequency. [c] Low frequency component of a doublet (see Fig. 16). [d] Very weak band. [e] Broad band. [f] Shoulder.

The bands corresponding to the C-halogen stretching mode proved useful for the identification of the guest conformation in the clathrates. The axial chair conformation of pure chlorocyclohexane absorbs at 685 cm^{-1} and this band persists with no frequency shift in the clathrate spectrum where it forms part of an asymmetrical doublet (Fig. 16). The less intense component (694 cm^{-1}) of the doublet is absent from the spectra of the other halogenated guest homologs and was tentatively assigned to the stretching vibration of the more strained C—Cl bond of the boat conformer. The overall appearance of the spectrum of TOT/bromocyclohexane clathrate resembles that of chlorocyclohexane (Table 7), thus suggesting the preferential inclusion of one axial conformation. Inspection of the IR spectrum of TOT/2-chlorotetrahydropyran demonstrates the same favored inclusion of an axial conformer in the clathrate. In fact, the guest exhibits the anomeric effect and the Cl-substituent occupies an axial orientation in the liquid-state as well [63]. Since the IR spectra of the three clathrates have most of their characteristic features in common, it was inferred that

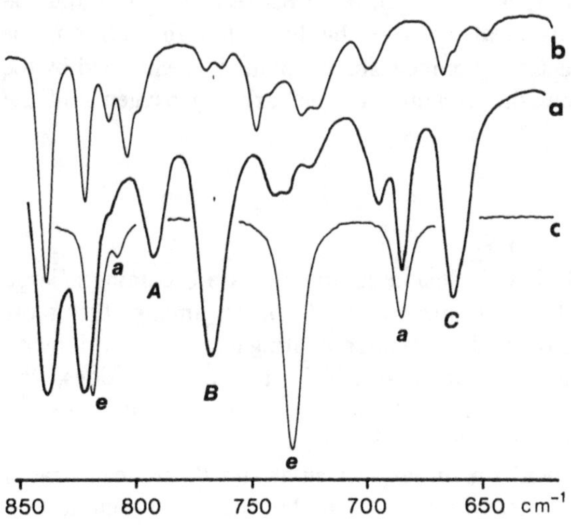

Fig. 16. General comparison of IR spectra: (a) TOT/chlorocyclohexane; (b) pure TOT; (c) pure guest. The characteristic bands assigned to the axial and equatorial guest isomers are denoted by **a** and **e** respectively. Bands **A**, **B** and **C** are usually observed in cage-type clathrates. Reproduced with permission from J. Incl. Phenom. *3*, 335 (1985)

the host-guest interactions are analogous from one clathrate to the other despite the differences in the crystal systems. Indeed, the three halogenated guest molecules are isosteric but the crystal morphology of TOT/2-chlorotetrahydropyran is unlike that of the usual cavity-type TOT inclusions. The clathrate crystallizes in a pseudohexagonal R system with axes $a = b = 17.142(3)$ and $c = 10.270(2)$ Å.

3.3.4 Inclusion of Monomeric Carboxylic Acids

Carboxylic acids exists in solution as equilibrium mixtures of monomers and hydrogen-bonded "polymeric" structures. Dimeric assemblies may even predominate in the gaseous phase [64]. Monomeric IR spectra of carboxylic acids have been recorded at 4K using argon or nitrogen matrix isolation techniques [65]. IR and Raman spectra have been reported for various β-quinol clathrates: the formic acid guest molecule vibrational frequencies were shown to be closer to the vapor-phase values than to the liquid- or solid-phase values and it has been suggested that formic acid is present as the monomeric species in the host lattice of β-quinol [66a]. More recently, the IR spectra of a series of Dianin clathrates containing homologous carboxylic acids have been interpreted in terms of a gradual changeover from dimeric formic acid to monomeric *n*-hexanoic acid [66b].

The isolation of monomeric acetic and propionic acids within cage-type TOT clathrates has been performed [22]. The IR spectra reveal sharp peaks at 3475 to 3445 cm^{-1} which disappear in the spectra of the acids mixed with unsolvated TOT. The strong C=O band of TOT overlap that of the enclathrated monomeric acid, however the strong band at 1701 cm^{-1} attributable to the hydrogen-bonded dimeric acid is clearly detectable in the mixture of the acid with unsolvated TOT. On crystallization of TOT from *n*-butyric and *n*-hexanoic acids, channel-type clathrates have been obtained [67]. The IR spectra in the range 3400–2300 cm^{-1} display wide bands characteristic of hydrogen-bonded dimeric aggregates, in accordance with unimpeded guest ··· guest contacts in the channels.

The crystal structure of the acetic acid clathrate corroborates the inclusion of a single molecule in the cage and reveals that only van der Waals host ··· guest contacts come into play [68]. There is no hydrogen bond between the guest and the TOT oxygens which might act as local hydrogen-binding sites. In addition, the significative "lock-and-key" character of the inclusion phenoma is exemplified by the absence of disorder resulting from the exchange of the carbonyl oxygen and the CH_3 group of acetic acid.

4 Concluding Remarks

Enclathration is not infrequent [69] and inclusion compounds are becoming a large family of well characterized chemical species. TOT ranks among the most attractive host molecules owing to the ability of differentiating between enantiomeric guest molecules in two different enantiomorphous host lattices. Besides, the unusually high propensity of TOT to form stable clathrates is further illustrated by no less than seven different achiral crystalline forms.

Attention has been mainly focused on the experimental and theoretical aspects of inclusion in chiral cage- and channel-type TOT clathrates, the former manifesting the highest enantioselectivity. A significant correlation between the absolute configuration of TOT and the preferentially included guest has been pointed out, however the ensuing feasibility for new guest configurational assignments has not been exploited to date. Possible reasons might be the limitations imposed by the guest size and by the experimental uncertainties arising from the low enantioselectivity of certain clathrates. Chiral discrimination in TOT clathrates has triggered a series of attempts to disclose the "nature" of the underlying host ··· guest interactions responsible for the phenomenon. Stereochemical considerations and packing energy calculations have provided models based essentially on X-ray structural results; whereas solid-state ^{13}C-NMR studies have led to models linked to the motional behavior of the guest molecules. At the present time the conclusions reached by the two approaches are somewhat at variance and further investigations, backed up by theoretical considerations, are called for to conciliate the two models. At this point it might be useful to note that the "rigidity" of the cavity has been recognized as an important factor in chiral discrimination, however the widespread occurrence of empty cavities in TOT clathrates and its influence on the overall chiral discrimination exerted by the host lattice do not seem to have been sufficiently appreciated.

The first cage-controlled assymetric reaction of an encaged prochiral olefin with an extraneous gaseous agent has been performed and should open up new vistas into the scarcely explored domain of crystal (diastereomeric) reaction paths. Another cage-controlled reaction has been provided by the *cis-trans* isomerization of two olefins whose reaction pathway proved to be different from that occuring in solution, or in the pure guest crystal. Symmetry relationship between the cavity and the included reactant are believed to be the main governing factors of these photoconversions.

The stable and selective inclusion of highly strained halocyclohexanes conformers or of monomeric carboxylic acids reveals the outstanding mutual fit of the TOT molecules constituting clathrate crystal, thus leading to a remarkable preservation of the cohesive van der Waals forces in the host lattice. But for a few exceptions, once formed the TOT clathrates prove to be thermally stable.

Two challenging features may be arbitratily retained from the present survey:

a) the ready availability of enantiomorphous TOT adducts should encourage further studies to improve the still debatable models intended for the understanding of guest chiral recognition;

b) the consideration of the host lattice cavities as rigid "chiral solvation shells" presents attractive potentialities as regards the initiation of asymmetric reactions on incorporated prochiral molecules. In addition, the inclusion of molecules in unusual conformations may be a source of new ideas on structural relationships between receptor and substrate.

5 References

1. Addadi, L., Ariel, S., Lahav, M., Leiserowitz, L., Popovitz-Biro, R., Tang, C. P.: Chemical Physics of Solids and their Surfaces, J. Chem. Soc. Special Periodical Reports *8*, 202 (1980)
2. a) Newman, A. C. D., Powell, H. M.: J. Chem. Soc. *1952*, 3747
   b) Lawton, D., Powell, H. M.: ibid. *1958*, 2339

3. MacNicol, D. D., McKendrick, J. J., Wilson, D. R.: Chem. Soc. Rev. *7*, 65 (1978)
4. Recent review: MacNicol, D. D.: Structure and design of inclusion compounds: the hexahosts and symmetry considerations, in: Inclusion Compounds, Vol. 2, (Atwood J. L., Davies, J. E. D., MacNicol, D. D., ed.) Academic Press, London 1984, p. 123
5. Backer, W., Harborne, J. B., Price, A. J., Rutt, A.: J. Chem. Soc. *1954*, 2042
6. Ollis, W. D., Stoddart, J. F. in: Inclusion Compounds, Vol. 2, p. 170 (Atwood, J. L., Davies, J. E. D., MacNicol, D. D. ed.) London Academic Press (1984)
7. Arad-Yellin, R., Green, B. S., Knossow, M., Tsoucaris, G.: 3rd International Symposium on Clathrate Compounds and Molecular Inclusion Phenomena, Tokyo, Japan (1984)
8. Farias, C. M., Hosangadi, B. D.: Indian J. Chem. *15 B*, 997 (1977)
9. a) Arad-Yellin, R., Green, B. S., Knossow, M., Tscoucaris, G.: J. Am. Chem. Soc. *105*, 4561 (1983)
   b) Arad-Yellin, R., Brunie, S., Green, B. S., Knossow, M., Tsoucaris, G: ibid. *101*, 7529 (1979)
10. Gerdil, R., Allemand, J.: unpublished results; Swiss National Science Foundation, Report 2.926-0.77 (1979)
11. Allemand, J., Gerdil, R.: Acta Crystallogr. *C39*, 260 (1983)
12. Gerdil, R., Frew, E.: J. Incl. Phenom. *3*, 335 (1985)
13. a) Allemand, J., Gerdil, R.: Cryst. Struct. Commun. *10*. 33 (1981)
    b) Allemand, J., Gerdil, R.: Acta Crystallogr. *B38*, 1473 (1982)
    c) Allemand, J., Gerdil, R.: ibid. *B38*, 2312 (1982)
14. Arad-Yellin, R., Green, B. S., Knossow, M., Tscoucaris, G.: Tetrahedron Lett. *1980*, 378
15. Williams, D. J., Lawton, D.: ibid. *1975*, 111
16. a) Brunie, S., Navaza, A., Tscoucaris, G., Declerq, J. P., Germain, J.: Acta Crystallogr. *B33*, 2645 (1977)
    b) Brunie, S., Tsoucaris, G.: Cryst. Struct. Commun. *3*, 481 (1974)
17. Ollis, W. D., Sutherland, I. O.: Chem. Comm. *1966*, 402
18. Gil, E., Quick, A., Williams, D. J.: Tetrahedron Lett. *1980*, 4207
19. Downing, A. P., Ollis, W. D., Sutherland, I. O., Mason, J., Mason, S. F.: J. Chem. Soc., Chem. Commun. *1968*, 329
20. Gerdil, R., Allemand, J.: Tetrahedron Lett. *1979*, 3499
21. Hamilton, W. C.: Acta Crystallogr. *18*, 502 (1965)
22. Arad Yellin, R., Green, B. S., Knossow, M., Rysanek, N., Tsoucaris, G.: J. Incl. Phenom. *3*, 317 (1985)
23. Gerdil, R., Allemand, J., Bernardinelli, G.: presented at the 12th International Congress of Crystallography, 12–25 August 1981, Ottawa, Canada.
24. Lee, B., Richard, F. M.: J. Mol. Biol. *55*, 379 (1971)
25. Kitaigorodsky, A. I. (ed.): Molecular Crystals and Molecules, Academic Press, New York 1973, p. 18
26. Powell, H. M.: Non-Stoichiometric Compounds, (Mandelcorn, L., ed.), Academic Press, New York 1964, p. 469
27. MaWu, N., Barrett, D. W., Koski, W. S.: Mol. Phys. *52*, 437 (1973)
28. Ripmeester, J. A., Burlinson, N. E.: J. Am. Chem. Soc. *107*, 3713 (1985)
29. Downing, A. P., Ollis, W. D., Sutherland, I. O.: J. Chem. Soc., B *1970*, 24
30. a)Andrew, E. R.: Int. Rev. Phys. Chem. *1*, 195 (1981)
    b) Schaefer, J., Chin, S. H., Weissman, S. I.: Macromol. *5*, 789 (1972)
31. Pines, A., Gibby, M. C., Waugh, J. S.: J. Chem. Phys. *59*, 509 (1973)
32. a) Ripmeester, J. A.: Chem. Phys. Lett. *74*, 536 (1980)
    b) Ripmeester, J. A., Tse, J. S., Davidson, D. W.: ibid. *86*, 428 (1982)
33. No intensity reported for this radical ion
34. a) Tabushi, I.: Acc. Chem. Res. *15*, 66 (1982)
    b) Wipff, G., Kollman, P. A., Lehn, J. M.: J. Mol. Struct. (Theochem) *93*, 153 (1983)
    c) Franke, J., Merz, T., Losensky, H.-W., Müller, W. M., Werner, U., Vögtle, F.: J. Incl. Phenom. *3*, 471 (1985)
35. a) Saenger, W., Noltemeyer, M., Manor, P. C., Hingerty, B., Klar, B.: Bioorg. Chem. *5*, 187 (1976)
    b) Review: Beddell, C. R.: Chem. Rev. *13*, 279 (1984), and references cited therein

36. Powell, H. M.: Nature *170*, 155 (1952)
37. a) Weinstein, S., Feibush, B., Gil-Av, E.: J. Chromatogr. *126*, 97 (1976)
    b) Shurig, V., Bürkle, W.: Angew. Chem. *90*, 132 (1978); Angew. Chem. Int. Ed. Engl. *17*, 132 (1978)
38. Arad-Yellin, R., Green, B. S., Knossow, M.: J. Am. Chem. Soc. *102*, 1157 (1980)
39. Gerdil, R., Allemand, J.: Helv. Chim. Acta *63*, 1750 (1980)
40. Individual values taken from ref. 9a
41. Williams, D. E.: Acta Crystallogr. *A 28*, 629 (1972)
42. Gerdil, R., Allemand, J., Bernardinelli, G.: ibid. *A 37*, C29 (1981)
43. Gerdil, R., Allemand, J.: TOT/(dl)-2-chlorobutane; unpublished X-ray structure (final $R = 0.056$ for 1429 independent reflections)
44. Gerdil, R., Bernardinelli, G.: TOT/(—)-1,2-dibromopropane; crystals grown from optically pure guest solution. Unpublished results
45. cf. Green, S. B., Heller, L.: Science *185*, 525 (1974)
46. For a recent general survey see: Origins of Life, Vol. 11; Special Issue: Generation and Amplification of Chirality in Chemical Systems (Thiemann, W., ed.), D. Reidel, Dordrecht, Holland 1981
47. Arad-Yellin, R., Green, B. S., Knossow, M.: Origin of Life (Wollman, Y., ed.), D. Reidel, Dordrecht, Holland 1981, p. 365
48. Wynberg, H.: Chimia *30*, 445 (1976)
49. Helmkamp, G. K., Joel, C. D., Sharman, H.: J. Org. Chem. *21*, 844 (1956)
50. a) Opella, S. J., Frey, H. M.: J. Am. Chem. Soc. *101*, 5854 (1979)
    b) Alemany, L. B., Grant, D. M., Alger, T. D., Pugmire, R. J.: ibid. *105*, 6697 (1983)
51. Goldup, A., Morrison, A. B.: Brit. Pat. 1.070.775 (C 07 c 7/02) (1967)
52. Farias, C. M., Hosangadi, B. D.: Indian J. Chem. *16B*, 1128 (1978)
53. a) Lamartine, R., Bertholon, G., Vincent-Falquet, M.-F., Perrin, R.: Seminaires Chim. Etat Sol. Vol. 10 (Suchet, J. P. ed.), Masson Ed., Paris 1976
    b) Thomas, J. M.: Pure Appl. Chem. *51*, 1065 (1979)
    c) Gavezzotti, A., Simonetta, M.: Chem. Rev. *82*, 1 (1982)
54. a) Miyata, M., Takemoto, K.: 3rd International Symp. on Clathrate Compounds and Molecular Inclusion Phenomena, Tokyo, Japan (1984)
    b) Miyata, M., Kitahara, Y., Osaki, Y., Takemoto, K.: J. Incl. Phenom. *2*, 391 (1984)
55. Gerdil, R., Barchietto, G., Jefford, C. W.: J. Am. Chem. Soc. *106*, 8004 (1984)
56. The photoconversions for *14* reported in ref. 22 are partially in error and have been corrected herein
57. Bregman, J., Osaki, K., Schmidt, G. M. J., Sonntag, F. I.: J. Chem. Soc. *1964*, 2021
58. Paul, I. C., Curtin, D. Y.: Acc. Chem. Res. *6*, 217 (1973)
59. Davies, J. E. D., Kemula, W., Powell, H. M., Smith, N. O.: J. Incl. Phenom. *1*, 3 (1983), and references cited therein
60. a) Allen, A., Fawcett, V., Long, D. A.: J. Raman Spectrosc. *4*, 285 (1976)
    b) Gustavsen, J. E., Klaeboe, P., Kvila, H.: Acta Chem. Scand. *A 32*, 25 (1978)
61. Allinger, N. L.: J. Am. Chem. Soc. *99*, 8127 (1977)
62. Rey-Lafon, M., Rouffi, C., Camiade, M., Forel, M. T.: J. Chim. Phys. *67*, 2030 (1970)
63. a) Booth, G. E., Ouelette, R. J.: J. Org. Chem. *31*, 544 (1966)
    b) Eliel, E. L.: Angew. Chem. *84*, 779 (1972); Angew. Chem. Int. Ed. Engl. *11*, 739 (1972)
64. Karle, J., Brockway, L. O.: J. Am. Chem. Soc. *67*, 898 (1945)
65. Berney, C. V., Redington, R. L.: J. Chem. Phys. *53*, 1713 (1970)
66. a) Davies, J. E. D.: J. Chem. Soc., Dalton Trans. *1972*, 1182
    b) Davies, J. E. D.: J. Incl. Phenom. *3*, 269 (1985)
67. Knossow, M., Tsoucaris, G.: personal communication
68. Arad-Yellin, R., Green, B. S., Knossow, M., Rysanek, N.: to be published
69. Andreetti, G. D.: J. Mol. Structure *75*, 129 (1981)

Structural Parsimony and Structural Variety Among Inclusion Complexes (with Particular Reference to the Inclusion Compounds of Trimesic Acid, N-(*p*-tolyl)-tetrachlorophthalimide, and the Heilbron "Complexes")

Frank H. Herbstein

Department of Chemistry, Technion — Israel Institute of Technology, Haifa, Israel 32000

Table of Contents

Topics in Current Chemistry, Vol. 140

1 Introduction

In this article we shall restrict ourselves to the discussion of those crystalline host-guest inclusion compounds, where the host forms a three-dimensional matrix in which the guest molecules are enclosed, either in channels or in cavities. The distinction between channels and cavities is often ill-defined because the presence of constricted regions in channels can give them partial cavity-like properties. We exclude those inclusion compounds in which the guest is contained in such host molecules like cyclodextrins [1], crown ethers [2] and cryptates [3]; the latter two types of hosts have been synthesized in the burgeoning activity of the past few years. Again the distinction may not be clearcut; in some cyclodextrins, for example, the guests are located between rather than within the host molecules (see discussion in Chapter 1 of this volume).

2 The Crystallisation Process

Let us start with a formalised over-simplified but convenient description of how organic crystals, their polymorphs and molecular aggregates are formed from solution. Crystallisation occurs when a suitable nucleus is present in a supersaturated solution. Consider first the formation of a pure crystalline phase — the molecules arrange themselves in the array of minimum free energy, assuming that equilibrium conditions are maintained. Directed forces, such as hydrogen bonds, may be the factor determining the form of the array, e.g. the densely packed tetragonal phase of urea. Polymorphism arises when two or more such arrays are possible, with only small differences in free energy among them; often kinetic rather than thermodynamic control prevails and a metastable polymorph forms first [4].

If the metastable polymorph has contact with a solvent then it reverts to the stable polymorph with the passage of time, but it may be indefinitely stable at room temperature under dry conditions. The enthalpy differences between polymorphs of organic crystals are generally only a few kJ/mol. Polymorphs of pure materials are generally close-packed i.e. without appreciable cavities in their structures. The density differences between polymorphs are usually about 1% or less.

A third possibility exists, which is especially relevant in the present context. Here the aggregation of lowest free energy is not the binary combination of pure crystal + separate solvent (or better saturated solution) but a new single-phase crystal containing both solute and solvent molecules, generally in fixed proportion. The situation which occurs in the formation of inclusion compounds can be depicted as follows. The host molecules form an array with cavities, which are empty in the hypothetical new phase. This phase is, in general, not thermodynamically stable because it is not close packed; this will be shown by a lower density calculated for the (hypothetical) host matrix on its own than for the stable polymorph [5]. However, the added interactions resulting from introduction of guest molecules into the cavities of the hypothetical structure are sufficient to stabilize the host-guest inclusion compound which constitutes a new binary phase in the phase diagram of the two-component system.

The interactions between host and host, and between host and guest (guest-guest interactions are generally negligible) may be due to van der Waals forces only, or to

Table 1. Various host molecules which form famililities of isostructural inclusion compounds. Distinction is made according to topological and interactive relationships of the host matrix

Matrix Topology	Matrix Interaction	
	Hydrogen Bonding	Van der Waals Bonding
Cage-type	Phenol [6] Quinol [6] Dianin's Compound [6]	Tetraphenylene [9] Triphenylmethane [16] Tri-*o*-thymotide [14]
Channel-type	Urea [7] Thiourea [7]	4,4'-Dinitrodiphenyl [8] Perhydrotriphenylene [15]

hydrogen bonding, or to a combination of the two (cf. Chapter I of this book). Thus, providing shape and size of the cavities are suitable, a *family* of inclusion compounds can be formed, with essentially the same host matrix but a variety of guests, whose chemical nature is not important as long as the geometrical requirements are fulfilled (Table 1). On the other hand, a specific type of host-guest interaction may place limitations on the chemical nature of the guests. If there are occupiable cavities in the neat crystals, then these may accept interstitial guests; such an aggregate would be a primary interstitial solid solution.

When the host molecule has a number of functionalities, or functional groups, permitting a variety of types of host-host or host-guest interaction, then one may anticipate that instead of a family of compounds individual structures will be formed, based on mutual adaptation of the bonding and steric requirements of host and guest. A range of intermediate situations is to be expected between these two extremes.

3 Some Families of Inclusion Compounds

Historically, emphasis was first directed towards families of inclusion compounds — the quinol clathrates [6] and the urea-paraffin hydrocarbon channel inclusion compounds [7] spring to mind. The host molecules are hydrogen bonded in these two examples, and there are only van der Waals interactions between host and guest. In the 4,4'-dinitrodiphenyl channel inclusion compounds [8] and the tetraphenylene clathrates [9] there are only van der Waals interactions between host molecules (and, indeed, between host and guest). An influence of the shape of the guest molecules on overall structure was demonstrated by Powell and Wetters [10] during the early years of crystallographic study of inclusion compounds: those formed by Dianin's compound (*1*) have cage structures when the guest molecules are small and globular but have channel structures with longer guests. Another set of early results suggested that an even greater variety of structural types would be found in the inclusion compounds of 2'-hydroxy-2,4,4,7,4'-pentamethylflavan (*2*) [11]; the crystal structures reported so far [12,13] show both resemblances and differences but a complete picture has not yet been obtained.

In this review we shall discuss in some detail three host molecules whose inclusion compounds display the range of possibilities mentioned above. The complexes of trimesic acid [TMA; (3)] demonstrate the relationship between polymorphism and complex formation in a rather complete way. This relationship is less apparent in the family of channel inclusion compounds formed by N-(*p*-tolyl)-tetrachlorophthalimide [TTP; (4)] but some hints can be discerned.

In the Heilbron inclusion compounds the possibilities of hydrogen bonding by the host molecule [DHDK; (5)] are such as to permit a rather varied mutual adaptation of the steric requirements of host and guest and thus a variety of different types of inclusion compounds are found rather than a single family or a group of families, as happens when TMA and TTP are the hosts.

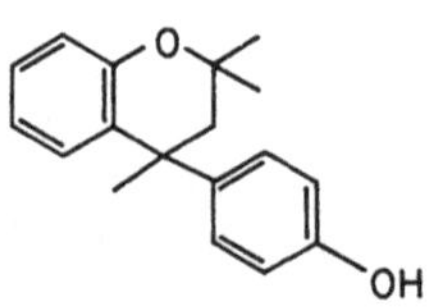

Dianin's compound, (1)

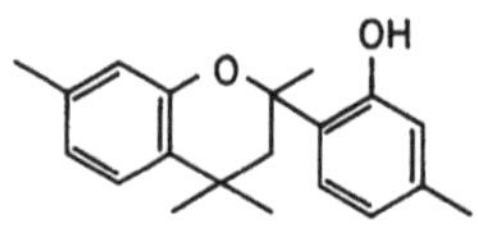

2'-hydroxy-2,4,4,7,4'-pentamethylflavan, (2)

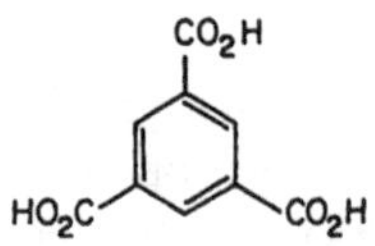

TMA, (3)

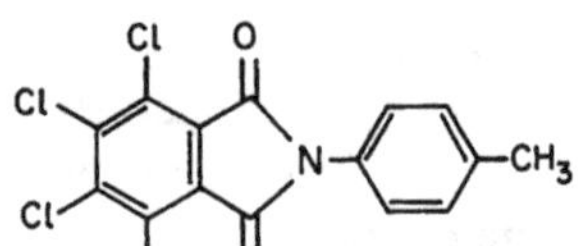

TTP, (4)

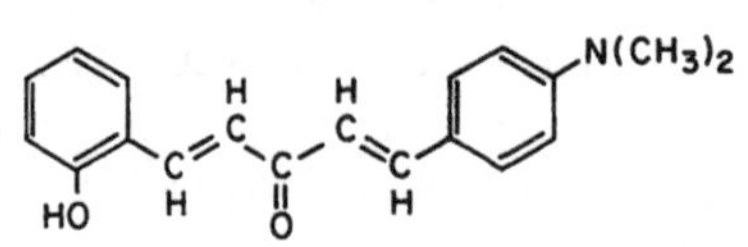

DHDK, (5)

4 Trimesic Acid (TMA) — Polymorphism and Inclusion Compounds

4.1 The Experimental Background

Trimesic acid (3) is a molecule with two major functionalities. Firstly it can act as a mono-, di- or tribasic acid and salts of all three dissociation steps have been reported. TMA can also form hydrogen-bonded cation-anion complexes with many amino acids and the structures of TMA · glycine monohydrate [17)] and histidinium trimesate · $^1/_3$ (acetone) [18)] have been reported. But these charged species are not of interest in the present context, as we shall concentrate attention on those complexes where the TMA molecule appears as an uncharged moiety exploiting to the maximum its possibilities of hydrogen-bond formation. The present description (see also Ref. [16)]) is intended to present facts and conclusions in logical rather than chronological fashion.

TMA behaves in a complicated way when crystallised from water, or from water containing various solutes, and the situation is made more puzzling by the tendency of different crystal types to appear together in the same crop under the rather uncontrolled conditions of temperature and concentration that generally prevail in

Table 2. Crystallographic results for the isomorphous α-TMA · xY interstitial clathrate compounds

Clathrate	α-TMA [19]	α-TMA · $^1/_6$ Br_2 [20]	α-TMA · $^1/_6$ Acetone [20]	α-TMA · $^1/_{12}$ I_2 [20]	α-TMA · $^1/_{12}$ Resorcinol [20]	α-TMA · $^1/_{12}$ Hydroquinone [20]	TMA · $^1/_{14}$ PA [20]
Space Group	*C*2/*c*	*C*2/*c*	*C*2/*c*	*C*2/*c*	*C*2/*c*	*C*2/*c*	*Ima*2 or *Imma*
a (Å)	26.520 (2)	26.510 (7)	26.541 (7)	26.541 (7)	26.577 (7)	26.531 (7)	19.52 (7)
b	16.420 (1)	16.449 (5)	16.482 (5)	16.457 (5)	16.434 (5)	16.450 (5)	127.6 (3)
c	26.551 (2)	26.580 (7)	26.604 (7)	26.530 (7)	26.585 (7)	26.559 (7)	16.53 (6)
β (deg.)	91.53 (1)	91.80 (1)	92.65 (1)	91.58 (1)	91.97 (5)	92.01 (5)	—
V (Å^3)	11558	11585	11625	11583	11596	11584	41172
Z	48	48	48	48	48	48	168
M_r	210.1	236.8	219.8	231.3	219.3	219.3	242.5
D_c (g cm^{-3})	1.45	1.63	1.51	1.59	1.51	1.51	1.53
D_m [a]	1.46	1.62	1.50	1.59	1.52	1.52	1.53

[a] Densities measured by flotation.

Table 3. Crystal data for the isostructural crystals of γ-TMA and three TMA polyhalides

	γ-TMA	TMA · I_5	TMA · IBr_2	TMA · Br_5
M	277.7	279.7	270.6	263.9
a (Å)	24.225 (7)	21.945 (7)	22.085 (7)	22.198 (7)
b	15.364 (5)	17.917 (6)	17.701 (6)	17.754 (6)
c	16.562 (6)	16.711 (6)	16.741 (6)	16.760 (6)
V (Å^3)	6164	6570.6	6544.5	6525.5
D_m (g cm^{-3})	1.47	1.72	1.64	1.62
D_c	1.47	1.72	1.65	1.61

1. There are 24 formula units per unit cell in all these crystals; 2. All the crystals have space group I222; 3. γ-TMA is actually TMA · 0.04 $C_6H_4(CO_2H)_2$ · 0.04 TMA

laboratory crystallisations. Fortunately the crystal habits are distinctive and the different types can be separated by hand under the microscope. The experimental results can be summarised as follows; the classification being based (by hindsight) on the structural information obtained in the later stages of the research:

(1) Equisided colourless cubes are obtained from water [19], water/acetone, and aqueous solutions of hydroquinone or resorcinol [20]. Brown cubes of similar habit are obtained from water saturated with Br_2, and purple cubes from water saturated with I_2 [21]. Yellow needles of composition α-TMA · $^1/_{14}$ PA were obtained from water containing picric acid (PA) [20]. The compositions and crystallographic results are summarised in Table 2.

(2) Reddish-brown needles (in various habits) are obtained from water containing KI and I_2 in equimolar amounts; analogous crystals are obtained from KBr—Br_2 and from KI—Br_2 solutions. The compositions are TMA · 0.7 H_2O · 0.09 HI_5 [21,

Table 4. Unit cell dimensions for TMA · H_2O · 2/9PA (I) and TMA · $^5/_6$ H_2O (II)

	I	II
a (Å)	18.269 (8)	16.640 (1)
b	8.852 (5)	18.548 (1)
c	3.642 (4)	9.512 (1)
α (deg)	90.43 (8)	95.81 (1)
β	92.59 (6)	91.06 (4)
γ	99.56 (6)	94.35 (1)
V (Å^3)	580.1 (7)	2911.3 (6)
Z	2	12
Space Group	$P\bar{1}$	$P1$

The net of cell I with edges [101] (= 18.47 Å), [011] (= 9.55 Å), inter-edge angle = 95.5° is very similar to the *bc* net of cell II. These two planes, (111) in cell I and (100) in cell II, contain the TMA · H_2O repeating network. The stacking of these nets along *c* (in cell I) and *a* (in cell II) differs in detail. The volume of cell II is 5.02 times as large as that of cell I, showing that the packing density of the layers is essentially the same in both compounds.

[23)], TMA · 0.7 H_2O · 0.103 HBr_5 [22)] and TMA · 0.7 H_2O · 0.167 $HIBr_2$ [21,23)] respectively. For convenience we shall refer to these compounds as TMA · I_5, TMA · Br_5 and TMA · IBr_2, respectively. The crystallographic results are summarised in Table 3.

(3) Colourless needles of composition TMA · 3 H_2O (which lose water on exposure to the atmosphere) and rhombic lozenges of composition TMA · $^5/_6$ H_2O (stable indefinitely) are also obtained from water. Yellow needles of composition TMA · H_2O · $^2/_9$ PA are obtained from water saturated with picric acid [22)]. The crystallographic results are summarised in Table 4.

(4) Needles of composition TMA · 2 H_2O are sometimes obtained from water [19)].

A further essential item of experimental information is that TMA undergoes apparent phase changes on heating (Fig. 1).

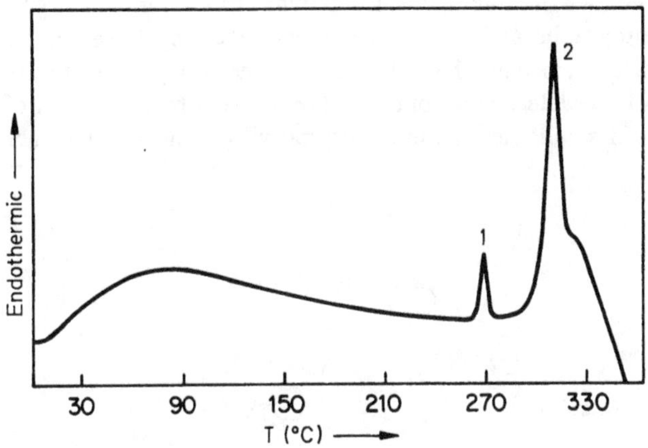

Fig. 1. Trimesic Acid (*3*): solid state reactions as recorded by DTA (heating rate 10 °C/min, in air). Together with parallel high-temperature X-ray diffraction studies, these results lead to identification of 1 as the α → β phase change and to 2 as the incompletely resolved combination of the β → γ quasi-phase change, vaporisation and melting (taken from Ref. [20)])

4.2 Deductions from Experiments

The complicated picture summarised above has been clarified by a series of single-crystal structure analyses extending now over a period of some sixteen years. Four structural families can be distinguished:

(1) α-TMA and the complexes listed in Table 2 [cf. Sect. 4.1 (1)] are isostructural interstitial primary solid solutions (clathrates). It is not yet entirely clear that TMA · $^1/_{14}$ PA [20)] belongs in this group.

(2) The high-temperature "polymorph" of TMA, hypothetical γ-TMA, and the polyhalide inclusions [cf. Sect. 4.1 (2); Table 2] are isostructural inclusion compounds. We shall show later that the polyhalide inclusions are channel-type compounds and that the so-called γ-polymorph is actually an interstitial (clathrate) inclusion compound with empty channels. Although γ-TMA is not a true polymorph in the strict thermodynamic sense, it is more convenient at this stage to be slightly inexact in nomenclature rather than pedantic.

(3) The examples listed in Table 4 [cf. Sect. 4.1 (3)] are isostructural channel inclusion compounds in which the essential structural element is TMA · H_2O and not TMA alone.
(4) The compound mentioned in Sect. 4.1 (4) is an OD crystal [25] of unknown structure and is the only example known so far in this group.

4.3 Polymorphism and Formation of Inclusion Compounds via Catenated[1] TMA Networks

The structure of α-TMA was determined by Duchamp and Marsh [19] in a *tour de force* in the days before diffractometers and Direct Methods (there are six TMA molecules in the asymmetric unit, and the intensities of 11563 reflections were measured by photographic methods). The two high-temperature polymorphs β and "γ" give similar Debye-Scherrer photographs and hence one infers that they have similar structures. β-TMA cannot be quenched and has a limited lifetime at high temperature so that its structure has not been determined, but "γ"-TMA grows by deposition of vapour on cooler surfaces and a suitable untwinned crystal was found for structure

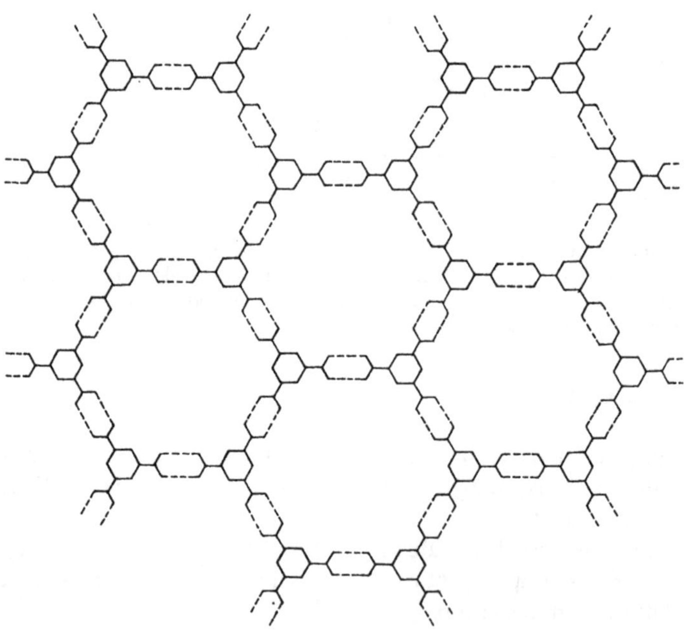

Fig. 2. The basic "chicken wire" motif in hydrogen-bonded TMA (*3*) is a two-dimensional network of six-molecule rings, the hydrogen bonds between carboxyl groups being represented by dashed lines. This network is planar in γ-TMA and folded (pleated) in α-TMA (taken from Ref. [19])

[1] A catenane or linked-chain structure (latin: *catena* = chain) is one in which macrocyclic molecules are held together mechanically without the aid of a chemical bond [26].

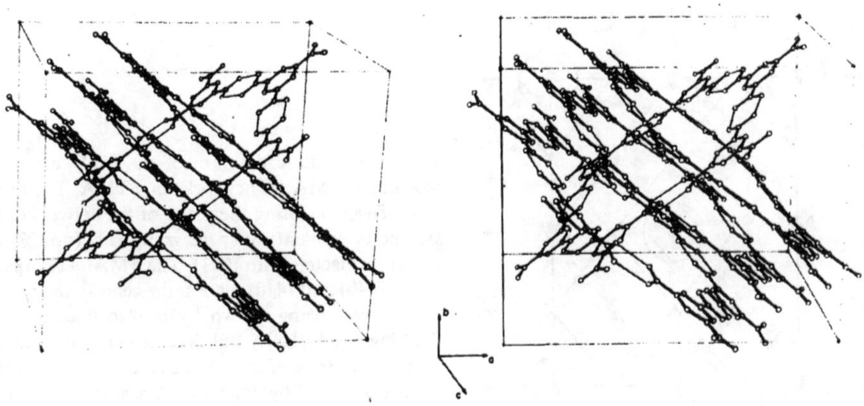

Fig. 3. The catenation of three hexagonal TMA (*3*) networks through a single such network; the two additional networks flanking the single network have been omitted for clarity. This stereodiagram* refers specifically to γ-TMA; the enclosing box has the dimensions and orientation of the unit cell but its origin has been shifted to 0,1/2,0 for aesthetic reasons. The same diagram with minor changes applies to the *local* situation in α-TMA (stereophotographs of a space-filling model are shown in Fig. 10, derived from Ref. [19])

* All the stereodiagrams in this paper were prepared using ORTEP [38]

analysis [20] some four years after the structures of the isostructural TMA polyhalides had been reported [23].

α- and γ-TMA are constructed in very similar ways in local or short-range terms but the long-range arrangements differ, in a way which is important for the phase relations between the two polymorphs. The common structural motif is the infinite hexagonal network of hydrogen-bonded TMA molecules and the remarkable feature resides in the way these networks are linked.

Consider the arrangement shown in Fig. 2. Each central hole with net diameter about 14 A (i.e. taking van der Waals radii into account) is threaded by three similar hexagonal arrangements, each parallel to the other (Fig. 3). Thus the hexagonal network with which we began is triply catenated by the other three interlaced networks. These in turn are catenated by the original network and by two other parallel networks on either side of the original one.

Thus a three-dimensional matrix is formed consisting of two sets of three parallel networks, each set being triply catenated by the other. While each network is internally hydrogen bonded, there is no hydrogen bonding between the three networks of a set, nor between one set and another. Each set of parallel networks is at an angle of about 116° to the second set. The descriptions now diverge for the two polymorphs and we shall begin with the simpler γ structure; when comparing the two structures it is useful to remember that the *b*-axis in α-TMA corresponds structurally to the *c*-axis in γ-TMA.

In γ-TMA the hexagonal networks (the "chicken wire") are all essentially planar and thus the two interlaced sets of parallel triplets cannot fill space efficiently but leave channels with axes along *c*. The overall arrangement is shown schematically

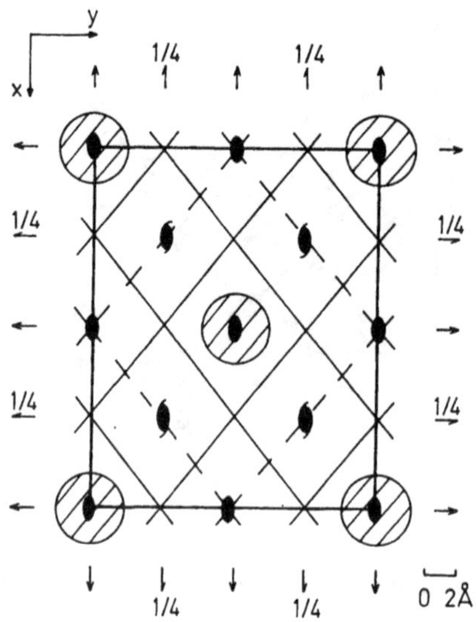

Fig. 4. Schematic diagram of the triply catenated hexagonal TMA (*3*) networks in γ-TMA. The direction of view is along the plane of the networks; the symmetry elements of space group I222 (no. 23) are shown projected onto (001). The TMA networks lie in the (440) and (4$\bar{4}$0) planes, the central network of each triplet being shown by broken lines and the flanking networks by full lines. The channels at the origin and centre of the diagram are empty in γ-TMA but are occupied by polyhalide chains in the isostructural TMA polyhalides. These polyhalide chains are viewed end-on and are represented by hatched circles (Taken from Ref. [23])

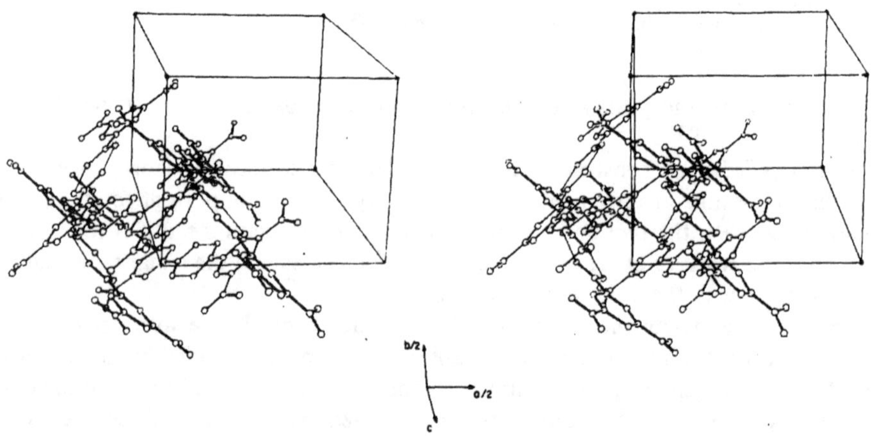

Fig. 5. Partial structure of γ-TMA (*3*), showing four portions of the TMA network (each comprising three TMA molecules) which intersect along the *c*-axis to form a channel of rhombic cross-section (angle of rhombus about 70 degrees). The linear polyhalide ions, which are not shown in the diagram, are contained in this channel. Only a quarter of the unit cell has been included in the diagram (taken from Ref. [23])

in Fig. 4, and in a stereodiagram of part of the unit cell in Fig. 5. This open arrangement of TMA networks would contain 24 TMA molecules per unit cell, with a calculated density of 1.36 g/cm^3. However the measured density is 1.47 g/cm^3 and our first guess (discreetly made in private after cell dimensions and densities had been measured but before the structure had been determined although its outline was known from that of TMA · I_5) was that the channels contained additional TMA

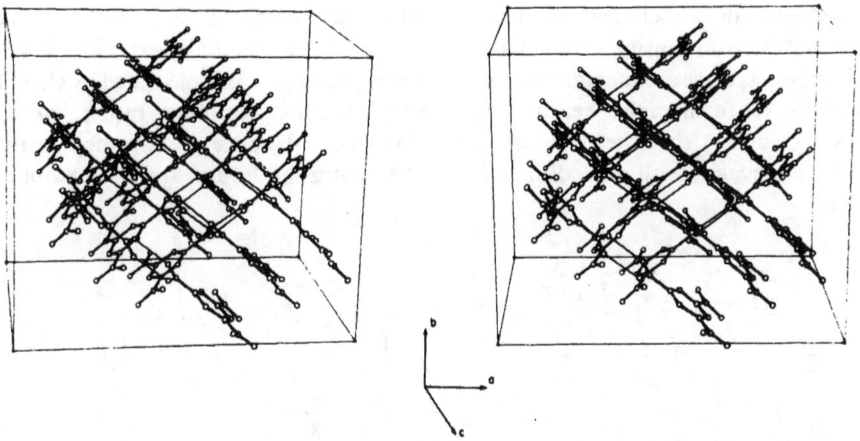

Fig. 6. Partial structure of γ-TMA (*3*), showing portions (each comprising four TMA molecules) of the interlaced triplets of TMA hexagonal rings which intersect about $^1/_2$, 0, $^1/_2$, leaving an open region. This appears to be occupied by disordered benzenedicarboxylic acid and TMA molecules, which are not shown. The origin of the unit cell has been shifted to 0, $^1/_2$, 0 for aesthetic reasons (taken from Ref. [23])

molecules. Detailed structure determination showed, however, that the channels were empty but that there was extra electron density in the cavities about $^1/_2$, 0, $^1/_2$ which are shown in Fig. 6.

The measured density can be neatly accounted for by assuming that these cavities contain about 4% benzenedicarboxylic acids plus 4% TMA. It has been shown that TMA decarboxylates at high temperatures [27]; our NMR analysis of a solution of "γ"-TMA showed an impurity content of about 4% benzenedicarboxylic acids. Our interpretation of these results is that the so-called "γ"-TMA is actually a "clathrate inclusion compound" stabilized by the presence of interstitial impurities. This is why we have been careful not to designate it as a true thermodynamic polymorph.

The impurities enclosed in the cavities cannot escape at high temperatures while escape would be easy under these conditions for any material contained in the channels. One may also speculate that the β-polymorph cannot be quenched because the stabilising impurities have not yet been produced by decarboxylation of TMA at the somewhat lower temperature of the α→β phase transformation. An intriguing possibility is to stabilise the "γ-polymorph" by growth from a solution containing, say, isophthalic acid with these molecules entering the channels as guests at room temperature. Our preliminary experiments in this direction have not yet been successful.

The channels along *c* are filled in TMA · I_5 and its analogues, the guest material being the appropriate polyhalide anions. These are disordered to a greater or lesser extent (depending on the nature of the anion) along the channel axes, showing that there are no loci of particular interaction between the TMA channel walls and the polyhalide anions. Where are the counterions to the polyhalide anions?

Although the crystals were grown from solutions containing potassium cations, chemical analysis shows that these are not incorporated. Thus the counterions must be protons; however the TMA framework is neutral and does not provide any possibility for attachment of protons. Difference syntheses for $TMA \cdot I_5$ show residual electron density in the cavities that contain the interstitial guests in "γ-TMA"; we infer that these are the water molecules whose presence is demonstrated by chemical analysis. We further infer, because of the absence of other possibilities rather than from direct evidence, that the protons needed for charge balance are attached to these water molecules, with related disorder of the neutral molecules and the hydronium ions.

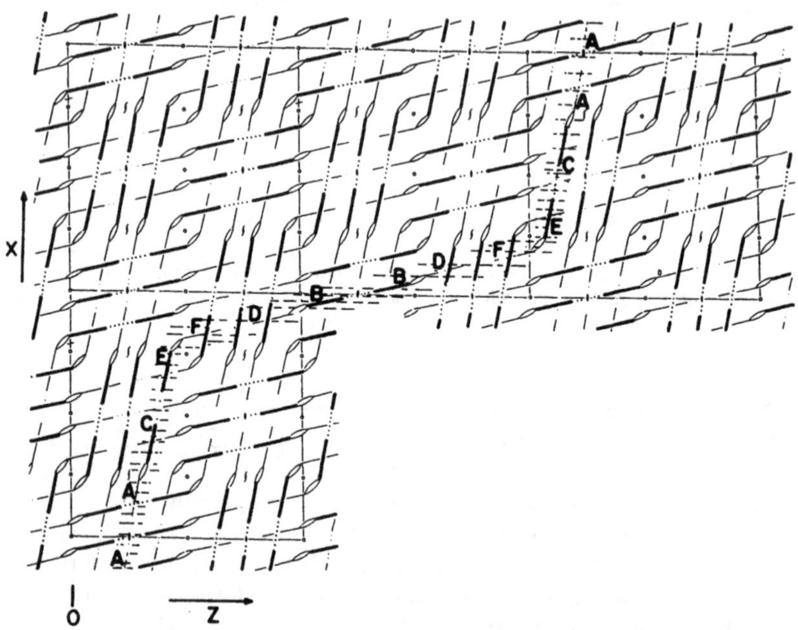

Fig. 7. Schematic representation of the α-TMA (*3*) structure viewed down the crystallographic b axis. Individual TMA molecules, seen edge-on, are represented by straight, solid lines. Dotted lines represent hydrogen bonds between molecules having similar *y* coordinates. Different heights along *y* are represented by different thicknesses for lines and dots. The pairs of curved lines represent hydrogen bonds between molecules lying at different heights along *y*. Four unit cells are shown and one network is emphasised to show how it progresses through the structure. The molecules are labelled to conform to Fig. 8 and to the labelling used in Ref. [19]. The molecules in a network are arranged as follows:

←---------------------- repeating unit ----------------------→

A A C E F D B B D F E C A A

---ce--- ---fl--- ---fl--- ---ce--- ---fl--- ---fl--- ---ce---

The network repeat unit contains twelve molecules. In the triply catenated interlacing of the various networks, a particular network takes up positions in the following order:

... ce fl fl ce fl fl ce ...

where ce represents a network *centred* between two others, which are termed *flanking* (fl).
This diagram is based on Fig. 6 of Ref. [19]

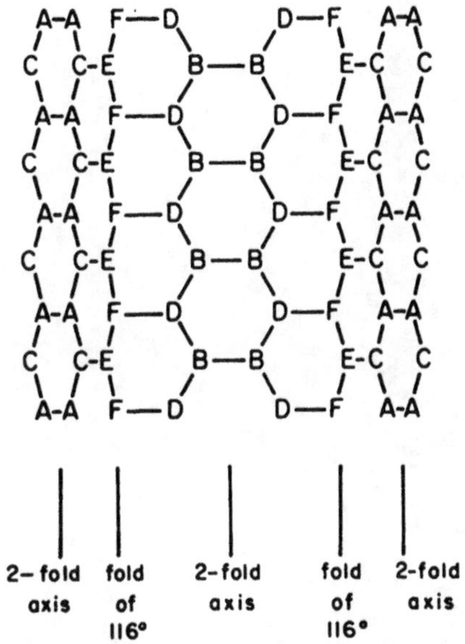

Fig. 8. α-TMA (*3*): structure of the network repeat unit emphasised in Fig. 7, shown here schematically in a view along [100]. The molecules F, D and B in this figure are essentially coplanar; however the hexagonal network undergoes an abrupt fold between F and E, with an angle of 116° between the mean planes (adapted from Ref. [39])

Some intriguing experiments can be proposed: K^+ is not incorporated into the crystal structure but what would happen to the smaller cations Na^+ and Li^+? Could thiocyanate replace polyhalide? Furthermore an analogue to the TMA · I_5 structure could be envisaged with neutral molecules, say straight-chain paraffins, replacing the anions in the channels. None of our preliminary experiments in these directions has yet been successful.

In α-TMA the networks are planar only for intervals of three hexagonal units, as

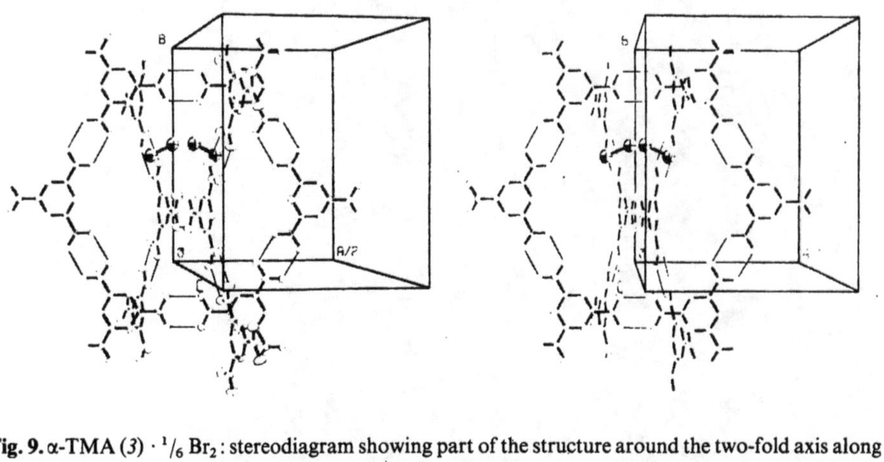

Fig. 9. α-TMA (*3*) · $^1/_6$ Br_2: stereodiagram showing part of the structure around the two-fold axis along 0, y, $^1/_4$. Only the central member of each of the two triple networks of hydrogen-bonded TMA molecules is shown, and only one of the disordered Br_2 orientations. The larger cavity referred to in the text is occupied by the Br_2 molecules while the smaller one is empty (taken from Ref. [20])

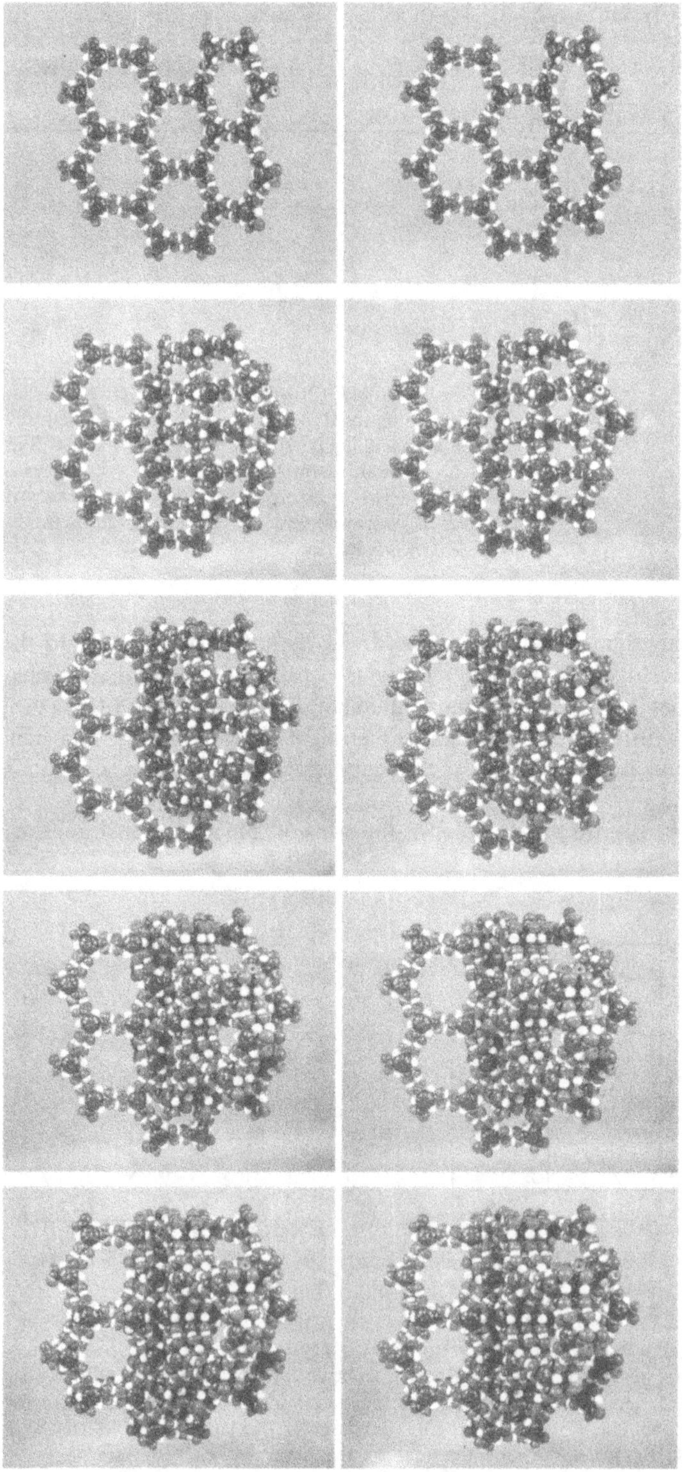

is shown in Figs. 7 and 8. Then, there is an abrupt change in the direction of the hexagonal network which arises from both rotation and displacement from the benzene ring planes of the hydrogen-bonded carboxyl groups joining the molecules denoted as E and F. This fold or pleat can only occur in the flanking hexagons of each catenated triplet and hence does not appear in the partial structure shown in the stereodiagram of Fig. 9, where only central hexagons are shown.

The line diagrams of Figs. 7, 8, and 9 develop in Fig. 10 which shows a series of striking stereophotographs of the α-TMA structure built up, network by catenated network, by Duchamp and Marsh [19] with space-filling models. The last of these stereophotographs shows that the *pleating* of the catenated TMA networks produces a rather efficient filling of space and channels are not formed. However two types of cavities remain in the region of the local triple catenation, one larger and one smaller (Fig. 9 and Fig. 10 (a)).

The interstitial Br_2 molecules of TMA · $^1/_6$ Br_2 are found in the larger of these voids in disordered fashion. Presumably the other guests listed in Table 2 are in the same set of cavities but this has not been tested directly. It was suggested that these cavities may contain water molecules even in the supposedly anhydrous α-TMA. However the density increment that would result is within the limits of experimental error, while the original intensity data are no longer available so that this suggestion cannot be subjected to experimental test without making a new set of intensity measurements.

The occurrence of analogous cavities in γ-TMA and TMA · I_5 has been noted above; however in these structures the two cavities are equivalent, being related by a crystallographic two-fold axis.

A consequence of the difference between the long-range arrangements of the catenated networks in α- and γ-TMA is that it is not possible to convert one arrangement into the other without breaking hydrogen bonds. The phase transformation (perhaps the α to β rather than the β to γ transformation) thus must be first order, as indeed it is found to be.

Interstitial primary solid solutions are rare among inclusion compounds but not unknown. Thus α-quinol contains small amounts of gas in solid solution, while Dianin's compound [6] forms inclusion compounds of primary solid solution-type, with only small differences in cell dimensions between neat crystals and interstitial solid solutions. The same situation occurs in α-TMA.

It is not yet clear how the TMA · $^1/_{14}$ PA structure should be classified but preliminary examination suggests that it is a modulated version of the α-TMA structure with picric acid molecules in interstitial rather than substitutional positions [the density of TMA · $^1/_{14}$ PA is higher than that of TMA itself (Table 2)]. Determination of the detailed structure would be a challenging project.

◀ **Fig. 10.** Stereophotographs of a space-filling model of part of the α-TMA (*3*) structure, showing how the three-dimensional structure is built up by triple catenation of two pleated "chicken-wire" TMA networks. The arrangement in (a) is directly comparable with the schematic diagram of Fig. 8; however the model comprises only that portion of Fig. 8 lying between the central and right-hand two-fold axes. (b) Two interlaced TMA networks. This part of the diagram is directly comparable to Fig. 9. (c) Three networks. (d) Four networks. This shows the interpenetration of one network by parallel portions of three other networks. (e) Six networks, showing the mutual interlacing of three parallel networks with three others in the second orientation. (Taken from Ref. [19])

The β-quinol clathrate inclusion compounds [6] are beautiful and early examples of solid-state catenation. In β-quinol each hydrogen-bonded network is single and three-dimensional; two such networks interpenetrate, to use Powell's original term, but are linked neither by covalent nor by hydrogen bonding. We would describe them as being catenated.

4.4 Inclusion Compounds Based on TMA · H_2O Networks

Let us now consider the hydrates, where the structural unit is found to be TMA · H_2O, polymorphism is not a factor and structural parsimony is neatly illustrated.

Although TMA · 3 H_2O has not been studied in detail, we can infer its structure by comparison with the other two complexes whose cell dimensions are given in Table 5. The TMA and H_2O molecules are linked in a planar framework containing channels of rectangular cross-section (Fig. 11). The remaining water molecules of TMA · 3 H_2O are presumed to be located in these channels but it is not known whether their positions are fixed along the channel axes; in any event escape is easy and the crystals decompose within a few hours on exposure to the atmosphere at room temperature.

Recrystallisation from water containing picric acid (PA) gives stable isomorphous yellowish needles, whose structure has been reported [24]. The TMA · H_2O framework of these crystals is well defined but the striking diffuse scattering indicates that the PA molecules located in the channels are partially disordered. This inability to position the guest with respect to the host matrix is a feature common to many channel inclusion complexes and makes it difficult to draw conclusions about the details of the host-guest interaction.

The TMA · $^5/_6$ H_2O structure provides a logical culmination to the other hydrate structures. The same planar TMA · H_2O sheets are present but here they are mutually shifted so as to generate a zigzag channel which can accomodate an ordered zigzag chain of TMA molecules hydrogen bonded through carboxyl groups in the meta positions, the third carboxyl of each molecule is not linked in any way (Fig. 12). Here the TMA molecules act both as hosts, through the TMA · H_2O networks, and as guests, within the channels. An analogous situation occurs in the segregated-stack charge transfer compound $(TMTTF)_{1.3}$ · $(TCNQ)_2$ [28], where the redundant TMTTF molecules, over and above those needed for the stacks, are located in channels between the stacks.

The overall structural formula for all these TMA · H_2O channel inclusion compounds can be written as

$$\mathrm{TMA \cdot H_2O \cdot [yX]}$$

where X is the guest (H_2O, PA or TMA in the present examples) and the value of y depends on the ratio of guest size to channel periodicity and has the following values:

$$y = 2 \quad \text{for} \quad X = H_2O$$
$$y = {}^2/_9 \quad \text{for} \quad X = \text{PA}$$
$$y = {}^1/_5 \quad \text{for} \quad X = \text{TMA}$$

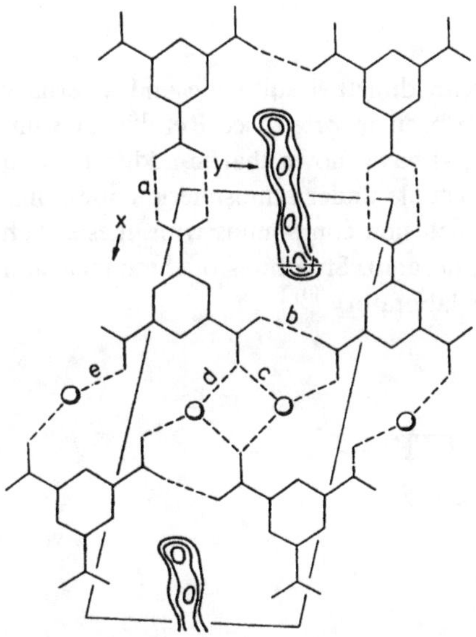

Fig. 11. Features of a difference electron density projection for TMA (*3*) · H_2O · $^2/_9$ PA. The difference electron density which represents the enclathrated picric acid molecules is contoured at levels of 1, 2, 3 e Å^{-3}. In TMA · H_2O · $^2/_9$ PA the molecules lie in the ($\bar{1}\bar{1}1$) planes and the hydrogen bonds do not lie in the plane of the projection but are directed out of this plane. For TMA · $^5/_6$ H_2O the present diagram serves as a slightly distorted representation of the arrangement in the layers of the framework TMA and H_2O molecules, as viewed along **a**. The hydrogen bonds have the following lengths, averaged over corresponding bonds in TMA · H_2O · $^2/_9$ PA and the five independent layers in TMA · $^5/_6$ H_2O (the values in parentheses are the RMS deviations from the means): a 2.641 (15) A; b 2.632 (52); c 2.874 (44); d 2.850 (52); e 2.592 (29). (Taken from Ref. [24])

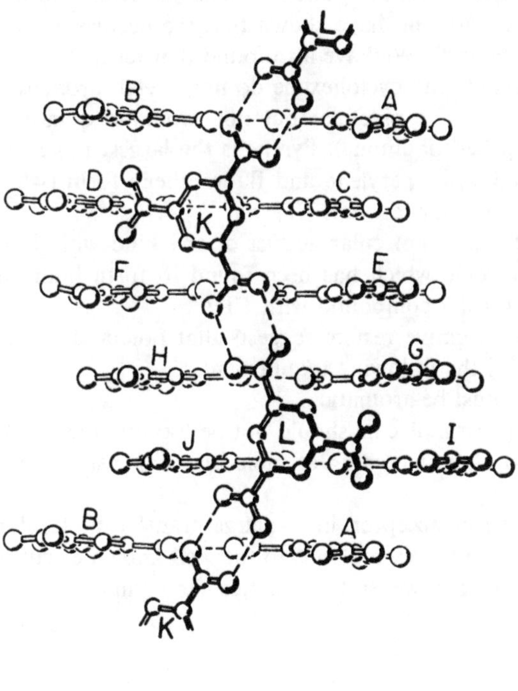

Fig. 12. The structure of TMA (*3*) · $^5/_6$ H_2O, showing the zigzag chains of clathrated molecules K and L and the zigzag stacking of the framework molecules A–J. All twelve TMA molecules are crystallographically independent. The view is along **c*** with a vertical (taken from Ref. [24])

4.5 Other TMA Inclusion Compounds

TMA forms a binary inclusion compound with dimethyl sulfoxide and a ternary with dioxane and water. Both structures have been reported (see Ref. [16] for summary). Kapon and Reisner (unpublished results) have shown that in addition to an inclusion compound with ethanol which is unstable under atmospheric conditions, there is a series of stable isostructural TMA inclusion compounds with guests such as iso-octane, n-tetradecane, n-octanol, and n-decanol. Structures of these inclusion compounds are now being determined in our laboratory [40].

5 N-(*p*-Tolyl)tetrachlorophthalimide (TTP) — Polymorphism and Inclusion Compounds

5.1 The Chemical Background

The chemical features of this system were established more than sixty years ago by Pratt and Perkins [29] who found that TTP (*4*) crystallizes from ethanol or glacial acetic acid as thin colourless plates but from aromatic solvents, such as benzene, toluene, the xylenes or from solutions containing aromatic molecules as long yellow needles.

It was recognised that the plates were neat TTP and the needles molecular adducts of composition 4 (TTP) · guest. Our own work [30,31] has confirmed and extended these results: we have found three polymorphic forms of TTP (Figs. 13 and 14), have confirmed that only guests with aromatic character participate in the formation of these adducts (see Table 6), and have shown that the needles are all isomorphous channel inclusion compounds. We have also found that tetralin forms an inclusion compound but that decalin and cyclohexane do not. Heteroaromatics are permissible guests (e.g. pyridine and α-picoline) but not aromatic quinones (e.g. *p*-benzoquinone and 2-methyl-*p*-benzoquinone). Pyrene is the largest molecule which forms an inclusion compound while perylene and fluoranthene form (what appear to be) π—π*-charge transfer compounds with TTP.

Tetrabenznaphthalene does not form a molecular adduct of any kind with TTP while phenothiazine is the only molecule which has been found to form both an inclusion compound and a charge transfer compound with TTP.

This combination of positive and negative results suggests that potential guests in the inclusion compound must fulfil the following requirements:

(1) Some part of the guest molecule must be aromatic

(2) The thickness and width of the guest molecule should not be too different from those of a benzene ring so as to allow matching of molecular dimensions with those of the channel.

The TTP molecule acts as an electron acceptor in π-charge transfer molecular compounds, as would be expected from its resemblance to the well-known electron acceptor tetrachlorophthalic anhydride; however these molecular compounds are not of interest in the present context.

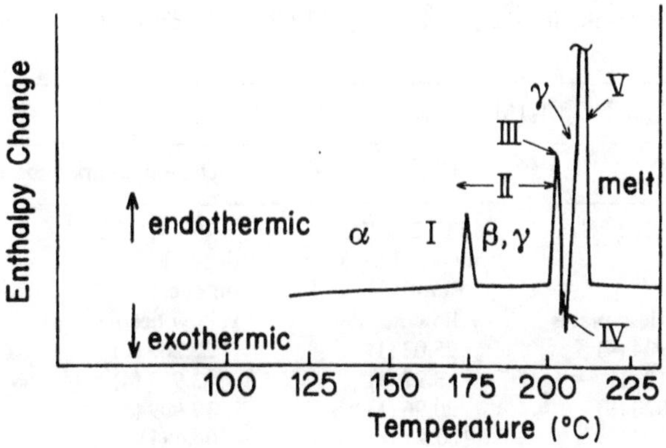

Fig. 13. DSC thermogram of α-TTP (*4*) heated from 25–225 °C (heating rate 8 °C min^{-1}). Temperatures of thermal events are defined as the beginnings of such events. The regions labelled refer to the following thermal reactions: I. Phase transformation $\alpha \rightleftharpoons \beta$ (reversible); II. Phase transformation $\beta \dashrightarrow \gamma$ (irreversible); III. Melting endotherm of untransformed β (i.e. not consumed in the $\beta \dashrightarrow \gamma$ transformation); IV. Solidification exotherm of melt $\rightarrow \gamma$; V. Melting endotherm of γ. The following transformation enthalpies (kJ $mole^{-1}$) were measured: $\alpha \dashrightarrow \beta$ 3.0 (2); $\gamma \dashrightarrow$ melt 19.5 (6) (taken from Ref. [31])

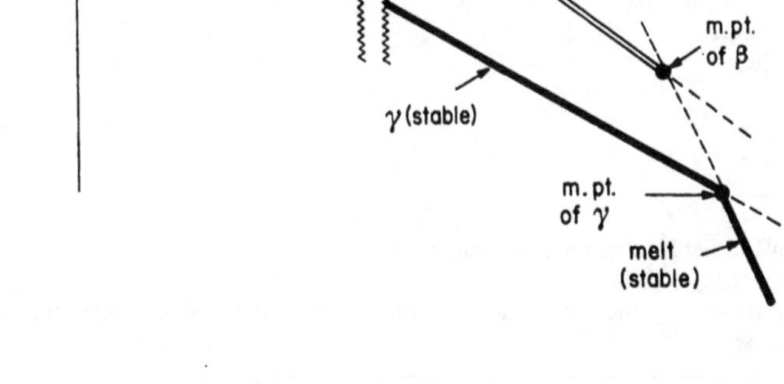

Fig. 14. Schematic free energy diagram for the polymorphs of TTP (*4*). The free energy scale is not specified as quantitative values are not available; the lines in the diagram are schematic and no attempt has been made to represent their positions or curvatures other than qualitatively. The transition points determined experimentally are shown as full circles; the $\alpha \rightleftharpoons \gamma$ transition is represented as an open circle as its temperature is not known accurately. The phases are denoted by filled lines for stable phases and open lines for metastable phases. Postulated extensions of free energy lines are shown as broken lines (taken from Ref. [31])

Table 5. Summary of crystallographic results for the polymorphs of TTP and its channel inclusion compounds

N-(*p*-Tolyl)tetrachlorophthalimide $C_{15}H_7NO_2Cl_4$ (M = 375.04)

	α-TTP	γ-TTP	Inclusion compounds
Stability range	<25°–173 °C	173°–214 °C metastable to 173 °C	<25°–150 °C depends on guest
Crystal form	colourless plates	yellow needles	yellow needles
a (Å)	6.904 (4)	25.02 (1)	22.267 (2)
b	25.19 (1)	3.97 (1)	3.911 (4)
c	17.020 (9)	30.96 (1)	19.909 (2)
β (deg)	—	100.9 (5)	106.4 (1)
V (Å^3)	2960 (3)	3020 (5)	1663 (1)
Space group	*Cmca*	$P2_1/c$	*C*2
Z	8	8	4

1. The β-phase is stable from 173° to about 200 °C; however it cannot be quenched and only the needle axis periodicity of 27.6 Å (approximately 4 × 6.9) was measured, at 174 °C; 2. Measurements for other phases at 25 °C; 3. The monoclinic subcell is given for the inclusion compounds (cf. text and Table 6).

Table 6. List of guest molecules used to prepare channel inclusion compounds of TTP, together with details of superstructure periodicity and overall symmetry of the crystals. This information is derived from the nature of the weak intermediate layer lines; if these are diffuse then the overall symmetry cannot be determined. The periodicity along the channel axis is given as a multiple (n) of the monoclinic repeat (3.91 Å). Results are for crystals at room temperature (25 °C)

Group I: sharp intermediate layer lines
n = 2 overall symmetry triclinic
bromobenzene, *o*-xylene
n = 3 monoclinic
anthracene, fluorene
n = 4 triclinic
picene
n = 7 monoclinic
tetralin, naphthalene, *o*-dichlorobenzene (polymorph I)
Group II: diffuse intermediate layer lines
n = 2 1-chloro-2-bromobenzene, *p*-xylene, α-picoline, *N*,*N*-dimethylaniline, 1,5-diaminonaphthalene
n = 3 phenothiazine, pyridine

1. Complexes with 1,2,5,6-dibenzanthracene and phenylisothiocyanate as guests have n = 4, but the very weak intermediate layer lines conform to monoclinic symmetry, contrary to the rule which appears to hold elsewhere that the symmetry is triclinic for n = 2, 4 and monoclinic for n = 3, 7; 2. Oscillation photographs showed no intermediate layer lines when the following guests were used: nitrobenzene, 1-chloro-3-bromo-benzene, *m*-xylene, 1,2,3,4-tetrahydroquinoline, azulene, 1-bromo-2-naphthylamine, biphenyl, *p*-toluidine, 9,10-dihydroanthracene, phenanthrene, pyrene; 3. All complexes have the composition 4:1 except for the picene complex where this ratio is 8:1.

5.2 Polymorphism of TTP

The crystallographic results for the polymorphs and complexes of TTP are summarised in Table 5 and 6. The structure of α-TTP is shown in Fig. 15. The dominant directional interaction is of the n ... σ type, between O and Cl, as shown by d(O ... Cl) = 2.98 A, while the antiparallel arrangement of adjacent molecules in the [001] direction suggests that dipole-dipole interactions are also important. The mean plane of the *p*-tolyl moiety is normal to the mean plane of the rest of the molecule, which lies in a crystallographic mirror plane. This perpendicular disposition of the mean planes of donor and acceptor parts of the molecule means that any intramolecular charge transfer absorption will be at high energies and the crystals are consequently colourless.

The β-phase of TTP is difficult to study; it is formed at 175 °C, according to DTA traces (Fig. 13) and high-temperature X-ray photographs. However it is not retained on quenching to room temperature and even at high temperature its lifetime is limited as it transforms to the γ-phase. There are some indications of a displacive transformation in which the original *a*-axis periodicity is multiplied by a factor of four; acenaphthylene shows somewhat similar behaviour [32].

The γ-phase has a number of special features. It is not formed within the bulk of the α-phase crystals (or perhaps of the β-phase after transformation from the α-phase) but the needles grow outwards from the original bulk crystals; a detailed study was

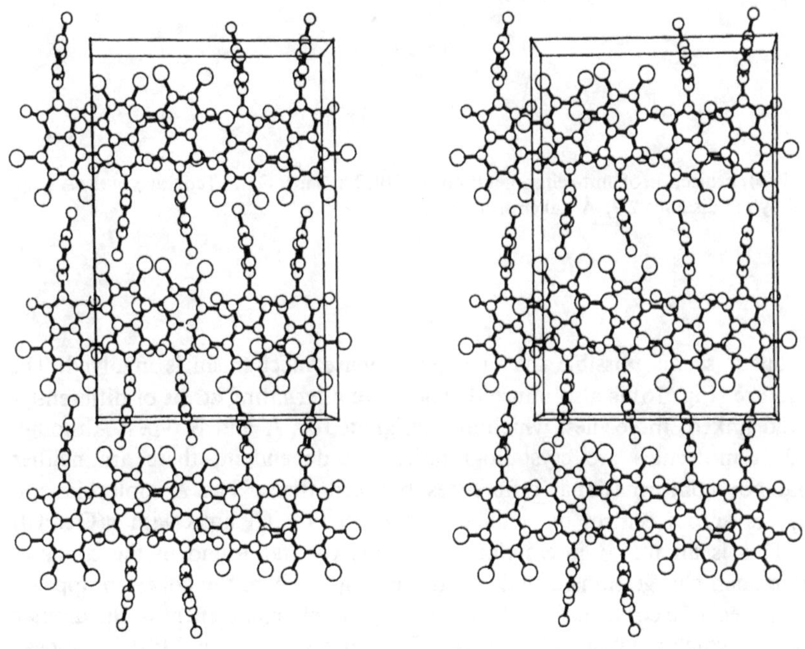

Fig. 15. Stereoview of the structure of α-TTP (*4*). The *a*-axis runs into the page, *b* is vertical and *c* horizontal (taken from Ref. [30])

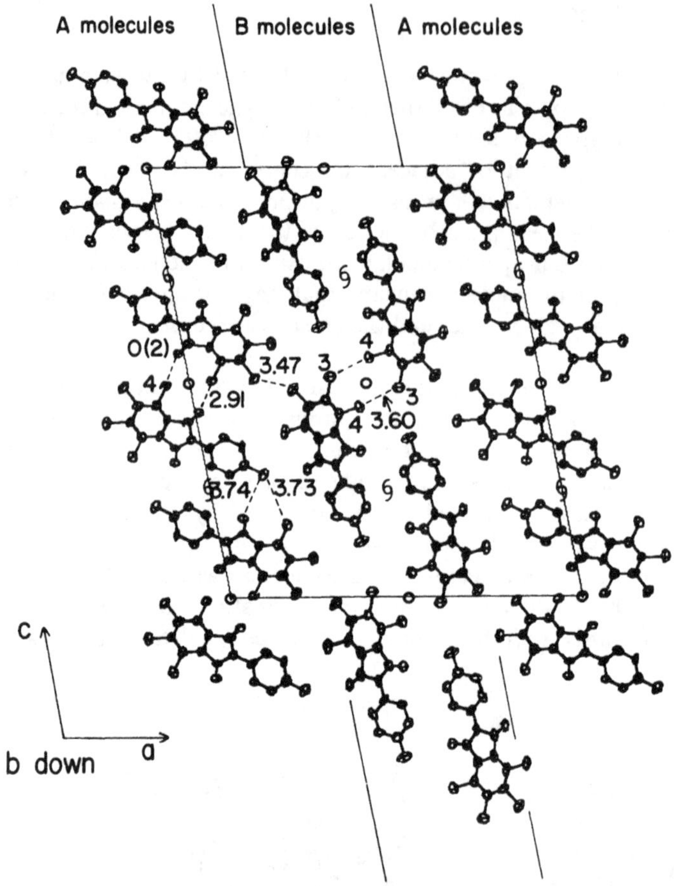

Fig. 16. γ-TTP (*4*): projection of unit-cell contents onto (010). The short O ... Cl contacts in the A strips are indicated by the distance 2.91 Å (taken from Ref. [31])

not made but it seems possible that a vapour phase mechanism is involved. The crystal structure (Fig. 16) is also unusual. There are alternating strips of differently-oriented (and linked) molecules, which are designated as A and B types. Although the A and B molecules are crystallographically independent, there are neither dimensional nor conformational differences between them. The A molecules are linked in pairs across centres of symmetry through O ... Cl links with d(O ... Cl) = 2.91 A. This is an n ... σ interaction analogous to that found in the α-phase. There are no such charge-transfer links in the B strips, where the cohesion appears to have an appreciable contribution from the antiparallel arrangement of the dipolar TTP molecules, which are related by centres of symmetry. The A and B strips appear to interact only by the usual van der Waals forces. In both A and B strips the molecules are stacked with stack axes parallel to the short *b* axis (3.97 Å).

5.3 Channel Inclusion Compounds of TTP

We shall first describe the structure of the isostructural inclusion compounds in overall terms and then return to the more subtle points. The structure (Fig. 17) is made up of stacks of antiparallel TTP molecules interacting only through van der Waals forces. The arrangement of molecules within the stacks is very similar to that in the γ-phase, as is shown by the close resemblance between the two *b*-axis periodicities (3.97 and 3.91 Å). The arrangement of the stacks is such that channels of ellipsoidal cross-section remain between them with the guest molecules contained in the channels; in fact the cross-section changes somewhat with progression along the channel axis. The walls of the channel are formed by four chlorines (two from each of two TTP molecules related by a crystallographic two-fold axis) and two oxygen atoms (one from each of two TTP molecules related to the first two by two-fold screw axes).

The TTP molecules are chiral because of the angle of about 55° between the planes of the tetrachlorophthalimide and *p*-tolyl parts of the molecule. As the space group C2 is chiral, spontaneous resolution has taken place on crystallisation but the absolute configuration of the crystal used in the structure determination could not be determined. It would be interesting to see if the chirality of the channel is adequate to resolve enantiomers but this was not attempted. The α- and γ-polymorphs crystalline in centrosymmetrical space groups and thus are racemates (in α-TTP the molecules lie on crystallographic mirror planes while in γ-TTP their conformation is similar to that found in the inclusion compounds). Of course the twisted TTP molecules will racemise rapidly in solution.

The cross-section of the channel is such that a benzene ring fits in rather easily and molecules such as anthracene, 1,2,5,6-dibenzanthracene and even pyrene can be accomodated; perylene and tetrabenznaphthalene are too large. Thus the size and shape selectivity are determined by the geometry of the channel, but this does not explain the requirement that the guests must have some aromatic character. Presumably this restriction arises from a special interaction between the π-systems of the aromatic guests and the electrophilic chlorines of the TTP molecules and might be expected to express itself in a particular location of the guest molecules with respect to the hosts along the channel axis. However, Herbstein and Kaftory [31] were unable to establish the location of the guests along the channel and thus some essential features of these inclusion compounds remain unexplained.

The cell dimensions of the various inclusion compounds do not vary with the nature of the guest. We have no explanation for this result which is rather surprising, especially when one remembers that there are only van der Waals interactions between the TTP molecules. On the other hand the symmetry of the unit cell does depend on the nature of the guest. Oscillation photographs of the inclusion complexes about their needle axes show a series of strong layer lines (corresponding to $b = 3.91$ Å) and another set of much weaker intermediate layer lines (sometimes diffuse or not discernible) which correspond to an integral multiple of 3.91 Å. Both host and guest molecules contribute to the strong layer lines, but only the guests contribute to the weaker layer lines.

Another unusual feature is that the symmetry of the subcell (inferred from measurements on the strong layer lines) is always monoclinic, whereas the overall

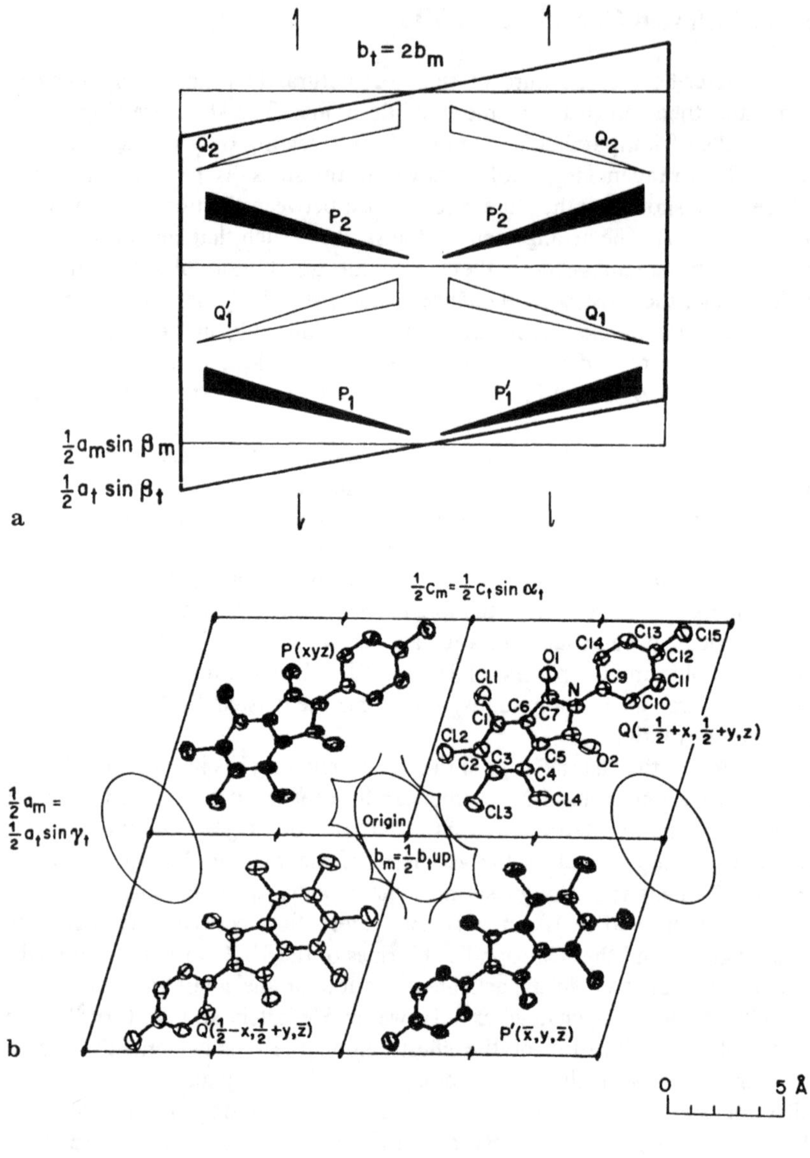

Fig. 17. 4 TTP (*4*). *o*-xylene inclusion compound.
a) Projection of unit cell down c_m ($= c_t$); two monoclinic cells are shown (lightly outlined) and one triclinic cell (heavily outlined). The eight crystallographically-independent TTP molecules in the triclinic cell are shown schematically, labelled so as to show their relationship to the four TTP molecules (of the monoclinic subcell) shown in Fig. 16b. **b**) Projection of unit cell down b_m; the monoclinic axes a_m, c_m are in the plane of the paper, as is c_t (because $\alpha_t = 90°$), but a_t runs down to the left below the plane of the page; the coordinates shown with the labels of the molecules refer to the monoclinic subcell. The van der Waals envelopes of Cl and O are indicated by ellipses but, as noted in the text, the actual cross-section of the channel depends on *y*. The molecules at *y* are shaded and those at $y + {}^1/_2$ open (taken from Ref. [31])

symmetry and size of the true cell (inferred from measurements on the whole diffraction pattern) depend on the nature of the guest. These experimental results are summarised in Table 6. It seems clear that these details hold the clue to the mutual disposition of host and guest sublattices but a model has not yet been proposed.

Another manifestation of the subtle influences exerted by different guests comes from the behaviour of the inclusion compounds on heating. They can be divided into two groups on this basis. In the first group (guests are, e.g., *o*-xylene, bromobenzene and, perhaps, tetralin) the decomposition occurs as follows:

$$4\,\mathrm{TTP.[guest]} \xrightarrow{\mathrm{guest}\uparrow} \gamma\text{-TTP} \xrightarrow{214\,^{\circ}\mathrm{C}} \mathrm{melt} \xrightarrow{178\,^{\circ}\mathrm{C}} \gamma\text{-TTP}$$

Representatives of the second group, where the guest include *o*-dichlorobenzene, *p*-xylene and *m*-xylene, decompose as follows:

$$4\,\mathrm{TTP.[guest]} \xrightarrow{\mathrm{guest}\uparrow} \alpha\text{-TTP} \quad ,$$

The α-TTP then behaves as described earlier (Fig. 13). The DTA results were confirmed by high-temperature Debye-Scherrer photography. However, no explanation was found for these differences in behaviour.

Our knowledge of the channel inclusion complexes of TTP has reached a stage of tantalising incompleteness. The overall structure of the complexes has been determined and appears rather simple and straightforward. However, the major chemical characteristic of these complexes, namely the aromatic nature of the guests, is quite without an established explanation while the structural basis for many unusual details of the diffraction patterns remains outside our grasp. It seems probable that only low-temperature diffraction studies will provide enough additional information to enable us to unravel these conundrums.

6 Molecular Inclusion Compounds of E,E-1-[*p*-Dimethylamino-phenyl]-5-[*o*-hydroxyphenyl]-penta-1,4-dien-3-one (DHDK) — the Heilbron "Complexes"

6.1 The Chemical Background

In 1921 Heilbron and Buck [33] reported that DHDK[2] (*5*) forms rather stable molecular "complexes" with 15 different guests whose chemical identities ranged from chloroform through phenanthrene (an electron donor) to 2,4,6-trinitrotoluene (TNT, an electron acceptor) (Table 7). The colours were reported to range from yellow to black, suggesting a variety of types of interaction between the components. Further-

[2] We give the currently accepted name for the compound in the title but use an acronym derived from the name used in Ref. [33], i.e. 4′-dimethylamino-2-hydroxydistyryl ketone.

Table 7. The list of guests which were reported to form molecular "complexes" with DHDK, according to the original study of Heilbron and Buck [33]

Addendum	Composition	Colour
Ethyl alcohol	$C_{19}H_{19}O_2N$, C_2H_6O	purplish-black
Ethyl acetate	$C_{19}H_{19}O_2N$, $^1/_2$ $C_4H_8O_2$	bluish-black
Benzene	$C_{19}H_{19}O_2N$, C_6H_6	almost black
	$C_{19}H_{19}O_2N$, $^1/_2$ C_6H_6	puce
Chloroform	$C_{19}H_{19}O_2N$, $CHCl_3$	bright red
		indigo-blue
m-Dinitrobenzene	$C_{19}H_{19}O_2N$, $C_6H_4(NO_2)_2$	reddish-violet
2,4,6-Trinitrotoluene	$C_{19}H_{19}O_2N$, 2 $C_7H_5(NO_2)_3$	reddish-brown
p-Benzoquinone	$C_{19}H_{19}O_2N$, $C_6H_4O_2$	greenish-black
Benzene and acetic acid	2 $C_{19}H_{19}O_2N$, $^1/_2$ C_6H_6, $^1/_2$ C_2H_4O	indigo-blue
Phenol	$C_{19}H_{19}O_2N$, C_6H_5OH	bluish-black
Phenanthrene	$C_{19}H_{19}O_2N$, $^1/_2$ $C_{14}H_{10}$	emerald green
Fluorene	$C_{19}H_{19}O_2N$, $^1/_2$ $C_{13}H_{10}$	bright green
Fluorene and methyl alcohol	$C_{19}H_{19}O_2N$, $^1/_2$ $C_{13}H_{10}$, CH_3OH	dark red
Carbazole	$C_{19}H_{19}O_2N$, $C_{12}H_9N$	crimson
p-Dimethylaminobenzaldehyde	$C_{19}H_{19}O_2N$, $C_9H_{11}ON$	brick-red
		light yellow

more they commented that "the ketone is very difficult to obtain in the free state as it tenaciously retains traces of solvent". In the only other report on this system [34] it was found possible to repeat some but not all of the Heilbron-Buck preparations and to determine the structures of three molecular complexes.

First let us note what could not be repeated: molecular adducts were not obtained with phenanthrene, fluorene, phenol, and carbazole. The other results were generally confirmed. Neat DHDK was only obtained in polycrystalline form since recrystallisation always gave an adduct containing solvent.

X-ray diffraction suggested that the (rather poor) crystals of the 1:2 TNT and TNB (1,3,5-trinitrobenzene) adducts were actually π-charge transfer compounds but that the other adducts were inclusion compounds. Determination of cell dimensions (Table 8) and comparison of Debye-Scherrer photographs suggested that the latter could be grouped as follows:

(1) The 2:1 inclusion compounds of DHDK with ethanol and chloroform (colourless form) were isomorphous, and these were isostructural with DHDK · CH_3COOH.

(2) The 1:1 inclusion compounds of DHDK with *m*-dinitrobenzene and anisaldehyde (4-methoxybenzaldehyde) appear to be related structurally.

(3) The only other inclusion compounds for which single crystals could be obtained were DHDK · CH_3OH and DHDK · PDMB (*p*-dimethylaminobenzaldehyde) and these appear to have different structures (cell dimensions in Table 8; structure of DHDK · PDMB below).

(4) Only polycrystalline samples were obtained of DHDK with $CHCl_3$ (blue form), CH_2Cl_2, aniline and hydroxylamine; these samples all gave different Debye-Scherrer patterns, which were also different from those of the other inclusion compounds noted above.

Thus, the only family of DHDK inclusion compounds which has been established so

Table 8. Single crystal data for some inclusion compounds of DHDK
DHDK = $C_{19}H_{19}O_2N$; PDMB = $C_9H_{11}ON$

Parameter	DHDK · 0.4 $CHCl_3$	DHDK · 0.5 C_2H_5OH	DHDK · CH_3COOH	DHDK · CH_3OH	DHDK · m-$C_6H_4(NO_2)_2$	DHDK · PDMB
a (Å)	12.086 (6)	12.039 (8)	12.379 (8)	13.931 (8)	21.787 (9)	22.383 (9)
b	10.323 (5)	10.252 (7)	10.194 (7)	12.541 (7)	13.850 (6)	12.272 (5)
c	8.015 (4)	7.879 (4)	7.819 (4)	10.570 (6)	7.759 (4)	8.917 (4)
α (deg)	94.58 (6)	93.64 (8)	94.78 (8)	68.28 (7)	88.25 (5)	—
β	103.58 (6)	104.64 (8)	96.70 (8)	73.33 (7)	84.70 (5)	92.99 (5)
γ	110.10 (6)	110.58 (8)	109.82 (8)	73.94 (7)	88.86 (5)	—
D_m (g cm^{-3})	1.27	1.21	1.27	1.32	1.31	1.21
D_c	1.27	1.21	1.28	1.34	1.31	1.20
Z	2	2	2	4	4	4
Space group	$P\bar{1}$[a]	$P\bar{1}$[a]	$P\bar{1}$[a]	$P\bar{1}$[a]	$P\bar{1}$	$P2_1/n$
V (Å^3)	898.6	868.3	913.9	1613.0	2329.9	2446.0

[a] Could also be $P1$.
1. The unit cells given for DHDK · 0.4 $CHCl_3$ and DHDK · 0.5 C_2H_5OH are both the Dirichlet and Delaunay reduced triclinic cells.
2. The unit cells given for DHDK · CH_3COOH and DHDK · m-dinitrobenzene are both Dirichlet (but not Delaunay) reduced cells.

far includes C_2H_5OH, $CHCl_3$ and (probably) CH_3COOH as guests. The others all have different structures, as is borne out by the results given below for DHDK · 0.4 $CHCl_3$, DHDK · *m*-dinitrobenzene, and DHDK · PDMB. These structures illustrate the mutual adaptation of host and guest molecules.

6.2 Comparison of the Structures

DHDK has an hydroxyl group as potential hydrogen-bond donor and a carbonyl group as potential hydrogen-bond acceptor. The only other functionalities which could potentially lead to formation of binary molecular compounds or "complexes" are the π-donor properties of the dimethylaminophenyl portion and the weaker π-acceptor properties of the hydroxyphenyl portion. Thus (with the advantages of some hindsight) we can envisage the following possibilities for formation of inclusion compounds:

(1) *Pairing of two DHDK molecules* by formation of two —OH ... O=C< hydrogen bonds. The resulting centrosymmetric dimer has a broad waist from which protrude long "tails" containing the *p*-dimethylaminophenyl moieties. The DHDK molecules are in the *s-trans, trans* conformation shown in formula *5* (see also Fig. 21). The overall shape of the dimer is not particularly conducive to close packing and thus voids are left between the molecule pairs which can be filled with guests such as $CHCl_3$ or C_2H_5OH. The structure of DHDK · 0.4 $CHCl_3$ is shown in Fig. 18 (as the ratio of DHDK molecules to cavities is 2:1, the experimental composition indicates that not all cavities are filled).

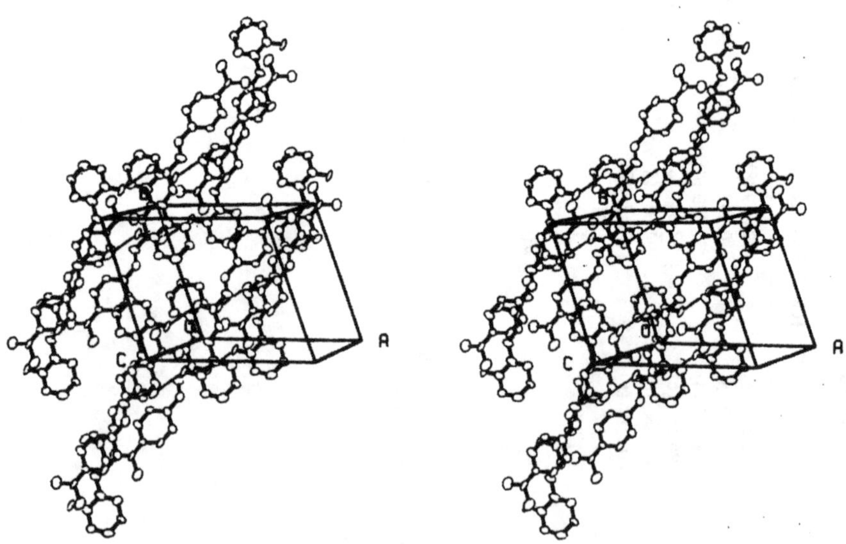

Fig. 18. DHDK (*5*) · 0.4 $CHCl_3$: stereodiagram of the crystal structure. The thermal motion ellipsoids represent 40% probability distributions (taken from Ref. [34])

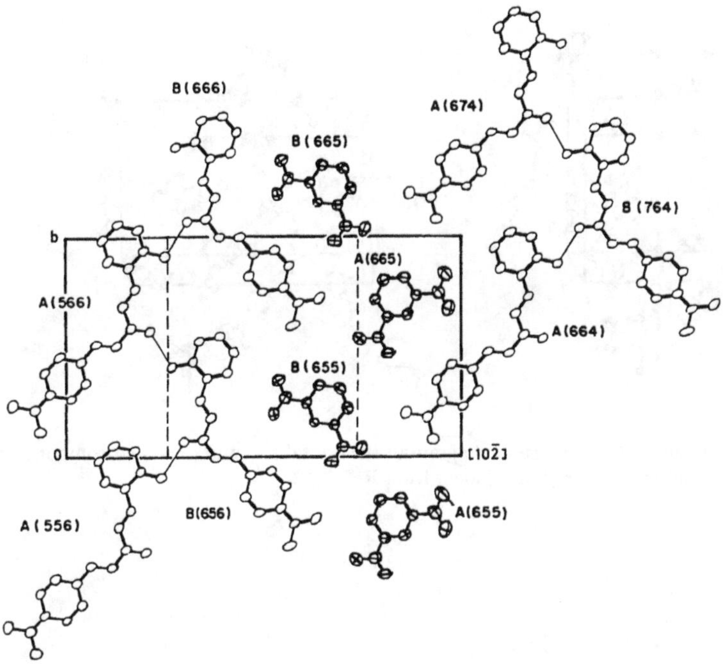

Fig. 19. The DHDK (*5*), *m*-dinitrobenzene structure, showing the sheet of molecules lying in the (201) planes. The two crystallographically independent molecules of each type are designated A and B. The reference molecules (coordinates tabulated in Table IV of Ref. [34)]) are denoted as 555, translations along the crystal axes being specified by adding or subtracting integers from the reference code, as in the ORTEP [38)] system. The rectangle shows the unit cell of the *pg* plane group, with its glide lines (taken from Ref. [34)])

This inclusion compound has a van der Waals bonded matrix and therefore is a "true" clathrate (cf. Chapter 1 of this volume); the hydrogen bonding is internal to the molecular pairs, and these interact with one another and with the guest by van der Waals forces only. A similar situation is found in the orthorhombic inclusion compounds of deoxycholic acid [35)] (see Fig. 5 in Chapter 1).

(2) *Formation of chains of DHDK molecules* linked by head-to-waist hydrogen bonds. The DHDK · *m*-dinitrobenzene inclusion compound is of this type (Figs. 19 and 20).

This structure is unusual in that it is based on the superimposition of binary sheets, such as that shown in Fig. 19. The DHDK molecules are arranged about non-crystallographic glide lines and the space left between them has a sinuous rather than linear shape. The *m*-dinitrobenzene molecules fit neatly into these wavy channels, the molecules also being arranged about non-crystallographic glide lines. There are only van der Waals interactions between the DHDK and *m*-dinitrobenzene molecules, and between the *m*-dinitrobenzene molecules themselves (true clathrate-type inclusion). The reference sheet has translationally-related similar sheets above and below it, but these are laterally shifted so as to

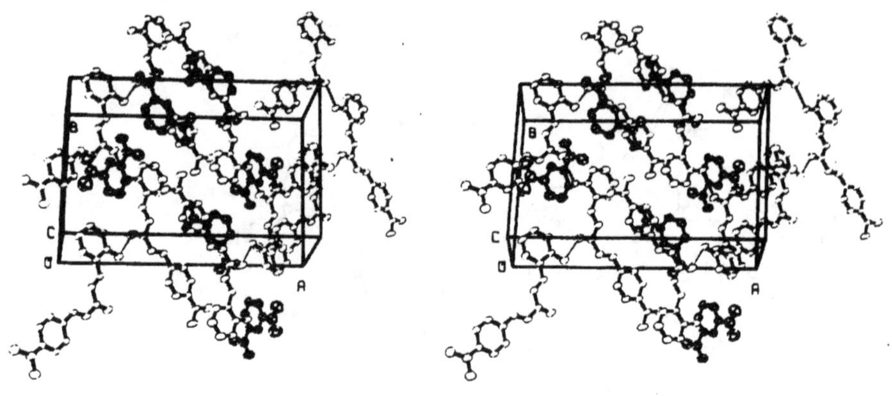

Fig. 20. DHDK (*5*). *m*-dinitrobenzene: stereodiagram of the crystal structure. The two components have been represented differently for clarity (taken from Ref. [34])

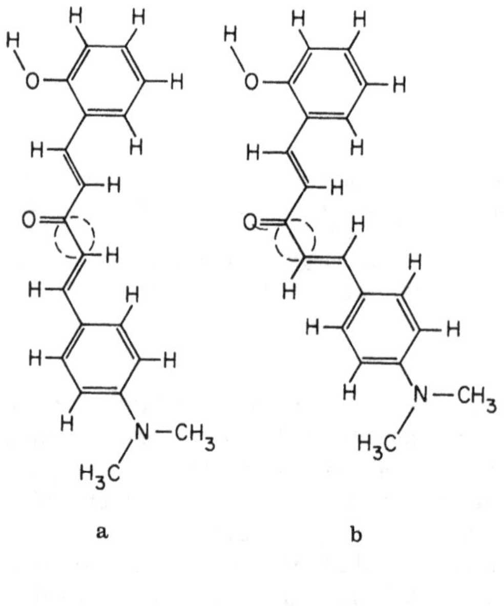

Fig. 21. Conformations of DHDK (*5*) molecules as found in some inclusion compounds: **a)** *s-trans,trans*, as in DHDK $\cdot$ 0.4 $CHCl_3$ and DHDK $\cdot$ PDMB; **b)** *s-cis,trans*, as in DHDK. *m*-dinitrobenzene. The single bond about which there is a difference in conformation is marked. (Taken from Ref. [34])

lead to enclosure of the *m*-dinitrobenzene molecules from above and below by strips of DHDK molecules (Fig. 20). The sheets interact only by van der Waals forces.

An unusual feature here is that the DHDK molecule in this aggregate has a different conformation from that found in DHDK $\cdot$ 0.4 $CHCl_3$; the *s-cis,trans* conformation allows the formation of pockets in the walls of the sinuous channels into which one of the nitro groups of each *m*-dinitrobenzene molecule fits neatly. The two molecular conformations are compared in Fig. 21.

(3) One must also take into account the possibility of *hydrogen bonding between DHDK and the guest* species forming a "Coordinatoclathrate"-type of inclusion com-

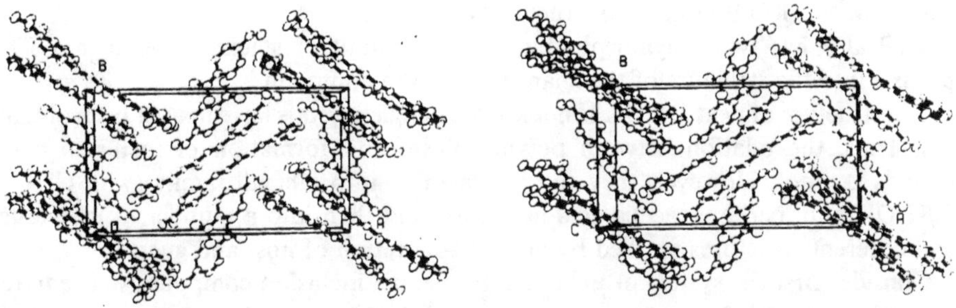

Fig. 22. DHDK (*5*) · PDMB: stereodiagram of the crystal structure (taken from Ref. [34])

pound (specification according to Chapter 1). This is illustrated by DHDK · *p*-dimethylaminobenzaldehyde where the hydroxyl of DHDK acts as hydrogen bond donor and the carbonyl of PDMB as acceptor.

The crystal structure of DHDK · PDMB (Fig. 22) has a centred array of DHDK molecules (in the *s-trans,trans* conformation, as in DHDK · 0.4 $CHCl_3$) which form the walls of channels; the PDMB molecules within these channels are hydrogen bonded to the DHDK host molecules. Between like molecules there are only van der Waals forces.

(4) *Formation of* π—π^* *donor-acceptor compounds* which are *perhaps* exemplified by the 1:2 molecular compounds of DHDK with TNB and TNT respectively. The caveat derives from lack of knowledge of the structures since the crystals obtained were of poor diffraction quality. If these are indeed π-molecular compounds, then the *p*-dimethylaminophenyl portion of DHDK would be expected to act as donor and the TNB or TNT molecule as acceptor.

Prediction of the structures of DHDK · *m*-dinitrobenzene and DHDK · PDMB would seem to demand either a clairvoyant structural chemist or an exceptionally powerful computer program.

7 Are There Other Systems Which Behave Similarly?

Despite the many differences among the three systems discussed above, they are nevertheless connected by relationships between polymorphism of the hosts and the formation of inclusion compounds. Do relationships of this kind extend to other systems already reported in the literature? Before answering this question, let us briefly consider the thread connecting the TMA, TTP and DHDK systems.

TMA has three polymorphic forms, the two known structures being characterised by an intricate hydrogen bonding scheme. α-TMA is closely packed but some cavities remain for the formation of interstitial inclusion compounds. γ-TMA is less closely packed and requires addition of interstitial guests for stabilisation of the quasi-polymorph; it can be further stabilised by incorporation of guests in the channels that remain empty in the quasi-polymorph. A different family of structures is formed when TMA · H_2O becomes the structural element.

TTP also has two polymorphs but only the motif of stacking found in γ-TTP persists into the structure of the channel inclusion compounds.

The structure of neat DHDK is not known (single crystals have not yet been obtained). Thus, the relation between polymorphism and formation of inclusion compounds is moot. However, they can be classified into a small family (with $CHCl_3$, C_2H_5OH and, perhaps, acetic acid as guests) and then into a number of individual and different structures marked by mutual adaptation of host and guest.

Consider first the system of urea and its channel inclusion compounds. The tetragonal framework of urea is so closely packed that interstitial inclusion compounds are not possible; the much more open hexagonal framework of urea molecules found in the channel inclusion compounds is only stable in the presence of guests of appropriate shape. There are resemblances to the TMA system which become closer when quinol is taken as host.

α-Quinol has a structure which is based in part on the cages formed by catenated hydrogen-bonded hexameric structural units [6,36]; this is another beautiful example of structural parsimony in nature. Depending on the method of preparation, α-quinol is either free of guests or can reach a composition of quinol: guest of 18:1. In β-quinol the whole crystal is composed of the interpenetrating hexamers, the more open framework being stabilised by the presence of the guests. Quinol does not form hydrates but the effect of the absence or presence of water is illustrated by the deoxycholic acid (DCA) system.

The structure of neat DCA does not appear to have been reported (it is difficult to obtain single crystals) but a wide range of isostructural orthorhombic channel inclusion compounds are formed in which the only hydrogen bonding is between hydroxyl and carboxyl groups of different host molecules [35]. When water is present in the crystals, then the hydrogen-bonding pattern changes markedly and tetragonal or hexagonal inclusion compounds are formed [37].

Thus, the relationship between polymorphism and formation of inclusion compounds seems to be widespread, although it can hardly be expected to be universal. Similarly, and not surprisingly, the presence of water (and presumably also of other molecules which can lead to formation of other hydrogen-bonding patterns) can have a marked effect on the type of inclusion compound that is formed.

The question of structural variety does not seem to be as widely explored and the DHDK system constitutes as yet perhaps the best example. However, there are a number of hints in the literature about other systems that would be worth studying.

8 Acknowledgements

The work reviewed here has been carried out at Technion and at Caltech. I am grateful to my co-workers R. E. Marsh (Caltech; whom I wish to thank specially for the original slides from which Fig. 10 was prepared), M. Kaftory, M. Kapon, G. M. Reisner, and M. B. Rubin (all at Technion). The work at Caltech was supported by various NIH grants to R. E. Marsh and that at Technion by the United States — Israel Binational Science Foundation (BSF), Jerusalem, The Fund for Basic Research of the Israel National Academy of Sciences, The Fund for the Promotion of Research at Technion and the Fund of the Vice-president for Research at Technion. I express my thanks for all this support.

9 References

1. Saenger, W., in: Atwood, J. L., Davies, J. E. D., MacNicol, D. D. (Eds.), Inclusion Compounds, Vol. 2, Academic Press, London, 1984, p. 8
2. Goldberg, I., in: See Lit. [1], p. 261
3. Dietrich, B., in: See Lit. [1], p. 337
4. Findlay, A.: The Phase Rule and Its Applications (updated by Campbell, A. N. and Smith, N. O.), Dover, New York, 1951[9], p. 54
5. Parsonage, N. G., Staveley, L. A. K., in: Atwood, J. L., Davies, J. E. D., MacNicol, D. D. (Eds.), Inclusion Compounds, Vol. 3, Academic Press, London, 1984
6. MacNicol, D. D., in: See Lit. [1], p. 1
7. Takemoto, K., Sonoda, N., in: See Lit. [1], p. 47
8. Saunder, D. H.: Proc. R. Soc. Lond., Ser. A *190*, 508 (1947)
9. Mak, T. C. W.: Clathrate Inclusion Complexes of Tetraphenylene, in: This Volume (1986)
10. Powell, H. M., Wetters, B. P. D.: Chem. Ind. (Lond.) *1951*, 266
11. Baker, W., Curtis,R. F., Edwards, M. G.: J. Chem. Soc. *1951*, 83
12. MacNicol, D. D., Mallinson, P. R.: J. Incl. Phenom. *1*, 169 (1983)
13. Gall, J. H., McCartney, M., MacNicol, D. D., Mallinson, P. R.: ibid. *3*, 421 (1983)
14. Ollis, W. D., Stoddard, J. F., in: See Lit. [1], p. 169
15. Farina, M., in: See Lit. [1], 69
16. Davies, J. E. D., Finocchiaro, P., Herbstein, F. H., in: See Lit. [1], p. 407
17. Herbstein, F. H., Kapon, M., Maor, I., Reisner, G. M.: Acta Crystallogr. B *37*, 136 (1981)
18. Herbstein, F. H., Kapon, M.: ibid. B *35*, 1614 (1979)
19. Duchamp, D. J., Marsh, R. E.: ibid. B *25*, 5 (1969)
20. Herbstein, F. H., Kapon, M., Reisner, G. M.: ibid. B *41*, 348 (1985)
21. Herbstein, F. H.: Isr. J. Chem. *6*, IVP (1968)
22. Kalina, D. W.: Ph. D. Thesis, Northwestern Univ., Evanston, Illinois 1979
23. Herbstein, F. H., Kapon, M., Reisner, G. M.: Proc. R. Soc. Lond. A *376*, 301 (1981)
24. Herbstein, F. H., Marsh, R. E.: Acta Crystallogr. B *33*, 2358 (1977)
25. Dornberger-Schiff, K.: Krist. Tech. *14*, 1027 (1979)
26. Schill, G.: Catenanes, Rotaxanes and Knots, Academic Press, New York, London, 1971
27. Shishmina, L. V., Belikhmaer, Ya. A., Sarymsakov, Sh., Koroleva, R. P., Sartova, K. A., Zhumanalieva, N. T.: Izv. Akad. Nauk Kirg. S.S.R. *1982*, 35 [Chem. Abstr. *98*, 106912n (1983)]
28. Kistenmacher, T. J., Phillips, T. E., Cowan, D. O., Ferraris, J. P., Bloch, A. N., Poehler, T. O.: Acta Crystallogr. B *32*, 539 (1972)
29. Pratt, D. S., Perkins, G. A.: J. Am. Chem. Soc. *40*, 198 (1918)
30. Kaftory, M.: Acta Crystallogr. B *34*, 471 (1978)
31. Herbstein, F. H., Kaftory, M.: Z. Kristallogr. *157*, 1 (1981)
32. Welberry, T. R.: Proc. R. Soc. Lond. A *334*, 19 (1973)
33. Heilbron, I. M., Buck, J. S.: J. Chem. Soc. *119*, 1500 (1921)
34. Herbstein, F. H., Kapon, M., Reisner, G. M., Rubin, M. B.: J. Incl. Phenom. *1*, 233 (1984)
35. Giglio, E., in: See Lit. [1], p. 207
36. Wallwork, S. C., Powell, H. M.: J. Chem. Soc., Perkin Trans. 2, *1980*, 641
37. Tang, C. P., Popovitz-Biro, R., Lahav, M., Leiserowitz, L.: Isr. J. Chem. *18*, 385 (1979)
38. Johnson, C. K.: ORTEP, Oak Ridge National Laboratory Report No. ORNL 3794, Oak Ridge, Tenn. 1965
39. Duchamp, D. J.: Ph. D. Thesis, California Institute of Technology, Pasadena, Cal. 1967
40. Herbstein, F. H., Kapon, M., Reisner, G. M.: J. Ind. Phenom. to be published (1987)

Inclusion Properties of Tetraphenylene and Synthesis of Its Derivatives

Thomas C. W. Mak and Henry N. C. Wong*

Department of Chemistry, The Chinese University of Hong Kong, Shatin, New Territories, Hong Kong.

Table of Contents

*Also known as N. Z. Huang

1 Introduction

This article reviews the history and synthesis of tetraphenylene (tetrabenzo[*a, c, e, g*] cyclooctatetraene), its ability to enclathrate a wide range of guest species, the structural characteristics of the resulting class of clathrate inclusion compounds, and synthetic approaches to the preparation of tetraphenylene derivatives which may serve as potential host systems.

2 Isolation of Tetraphenylene and Its Structural Characterization

The unexpected isolation of tetraphenylene *1* was claimed in 1928 [1]. However, careful determination of the molecular formula by elemental analysis revealed that this sample was actually *p*-quaterphenyl *2* [2].

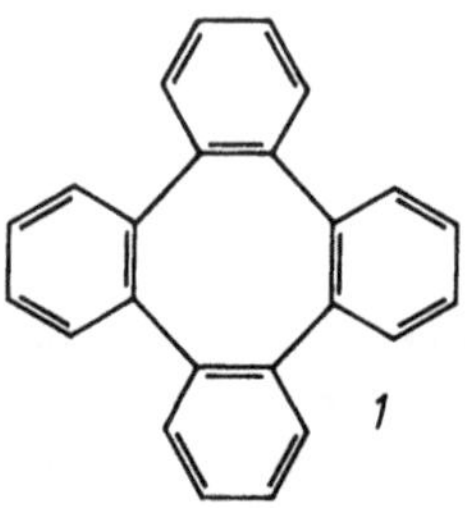

1

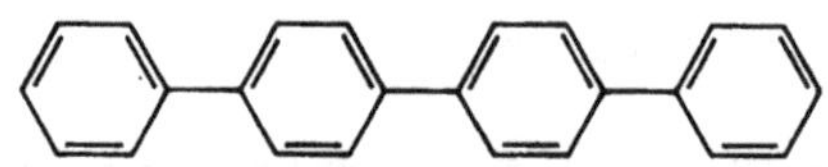

2

The first successful synthesis of tetraphenylene *1* was reported in 1943 by Rapson, Shuttleworth and van Niekerk, [3] who converted 2,2′-dibromobiphenyl *3* to its corresponding Grignard reagent *4*. The addition of copper(II) chloride to *4* provided tetraphenylene *1* in only 16% yield, together with 4% of biphenylene *5* as a minor product (Eq. 1).

(1)

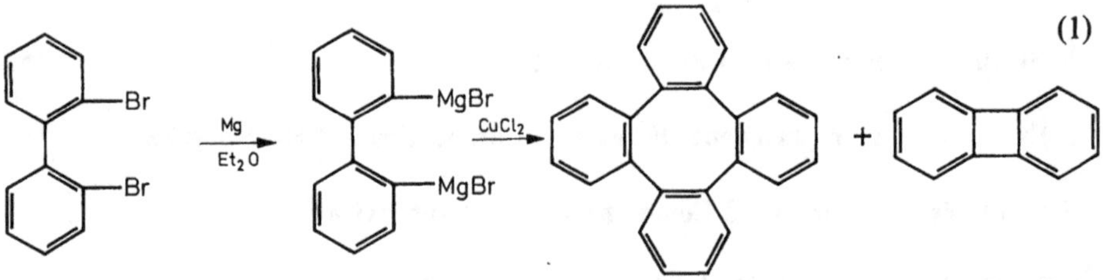

3 *4* *1* (16%) *5* (4%)

Tetraphenylene *1* forms colorless, high-melting (m.p. 233 °C) and exceedingly stable crystals. Its solubility in most solvents is relatively low. Nevertheless, *1* was observed to afford solvated forms as transparent rectangular prisms from benzene, carbon tetrachloride, dioxan, pyridine, chloroform as well as acetone [3]. The empirical molecular formulae of these solvates were established as $2\ C_{24}H_{16} \cdot G$ where G = solvent molecule [3].

Rapson et al. reported the unit-cell data for *1* (a = 15.60, b = 13.16, c = 16.39 Å, β = 100° 40′, space group $C2/c$, D_m = 1.20—1.25 g cm^{-3}, and Z = 8) and stated that its crystal structure was under investigation [3], but the purported study was apparently unsuccessful as no follow-up account ever appeared. The molecular structure of *1* was established by electron diffraction in 1944 [4]. It was shown that the central cyclooctatetraene ring of *1* is tub-shaped with bonds alternating around the ring, the shorter bonds being common to the benzene rings. Furthermore, the benzene rings are disposed alternately above and below the mean plane of the molecule. The absorption spectra [5] of *1* in hexane, ethanol as well as cyclohexane also indicated that the central cyclooctatetraene ring of *1* was not planar and it contributed very little to the resonance of the molecule. The calculations based on a non-planar model of *1* also provided transition energies in good agreement with the experimental data [6]. Finer details of the ground-state geometry of *1* was finally determined by an X-ray crystallographic study in 1981 [7].

3 Synthesis of Tetraphenylene

Although successful preparations of tetraphenylene *1* were reported by several research groups, some of these methods only involved isolation of extremely low yields of *1* from very complex reaction mixtures. Indeed, very few methods are available for the simple synthesis of tetraphenylene *1* in high yield. As some of the following examples will demonstrate, the isolation of *1* often required rather tedious and painstaking procedures.

o-Phenylenemercury *7*, generated in 52% yield by treatment of *o*-dibromobenzene *6* with sodium amalgam, was converted to *o*-dilithiobenzene *8* through lithiation (Eq. 2) [8]. When *o*-dilithiobenzene *8* was quenched by the addition of nickel(II)

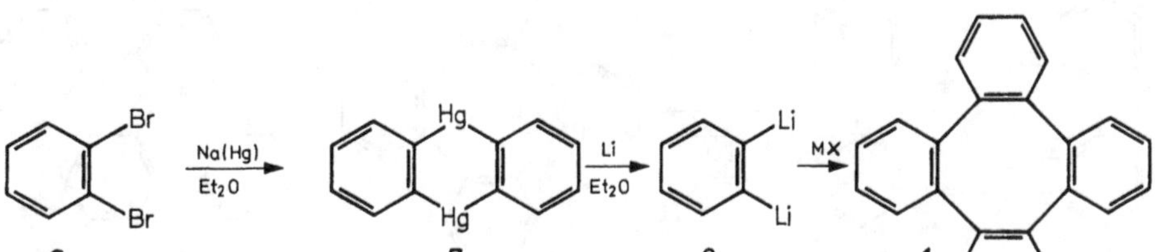

6 *7* *8* *1*

MX	Yield of *1* (%)
$NiCl_2$	11
$CuCl_2$	7
CuCl	9
VCl_2	5
$PdCl_2$	9
AgCl	3

(2)

chloride, copper(II) chloride, copper(I) chloride, vanadium(II) chloride, palladium(II) chloride, as well as silver(I) chloride, tetraphenylene *1* was isolated in 3—11% yield together with a mixture of other products [8,9].

On the other hand, 2,2′-dilithiobiphenyl *10*, prepared from 2,2′-diiodobiphenyl *9* through lithiation, could be converted to tetraphenylene *1* in 21—53% yield by metal chlorides such as nickel(II) chloride, copper(II) chloride, copper(I) chloride, copper(II) chloride-bipyridyl complex, manganese(II) chloride, iron(III) chloride,

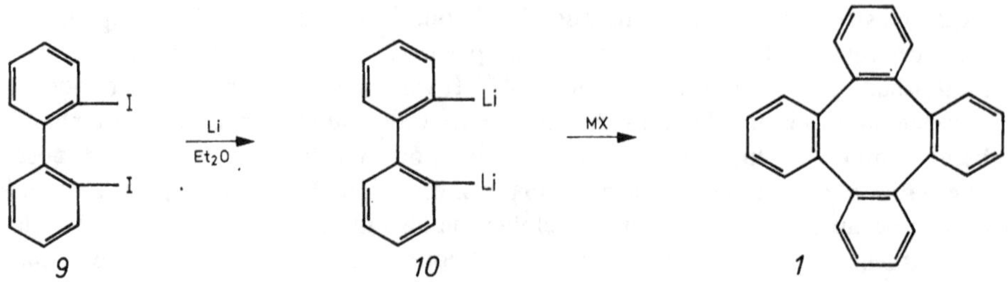

MX	Yield of *1* (%)
$NiCl_2$	44
$CuCl_2$	53
CuCl	21
$CuCl_2 \cdot$ bipy	47
$MnCl_2$	22
$FeCl_2$	47
$MoCl_5$	41

(3)

molybdenum(V) chloride (Eq. 3) [10,11]. Again it was necessary to isolate tetraphenylene *1* from complex mixtures of many side products. The possible function of copper(II) chloride and the mechanism of this reaction have been comprehensively discussed in a review [12].

It was also recorded that when a mixture of 2,2′-diiodobiphenyl *9* (3.6 g) and 2-iodobiphenyl *11* (12.4 g) was coupled by treatment with butyllithium and cobalt(II) chloride in *n*-heptane, tetraphenylene *1* (0.15 g) was isolated together with *o*-quaterphenyl *12* and *o*-sexiphenyl *13* in 26.6% and 5.7% yields, respectively [13].

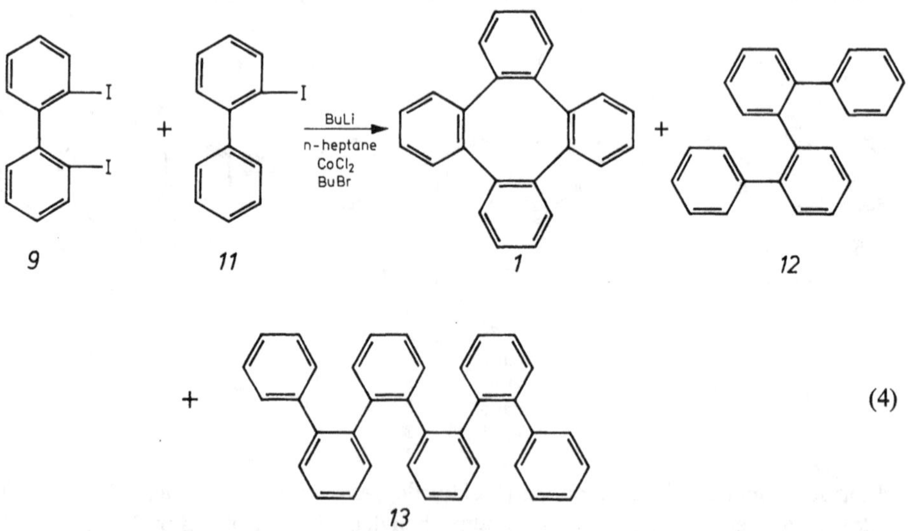

(4)

When a solution of biphenylene *5* in benzene was allowed to react with $[Ni(CO)_2(PPh_3)_2]$ in a sealed tube at 100 °C for 7 hours, 10% yield of tetraphenylene *1* was isolated (Eq. 5). The dimerization process was believed to proceed through the formation of a biphenylene-nickel complex [14].

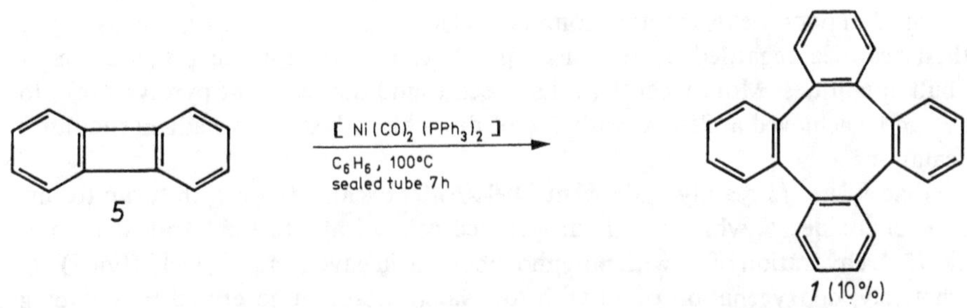

(5)

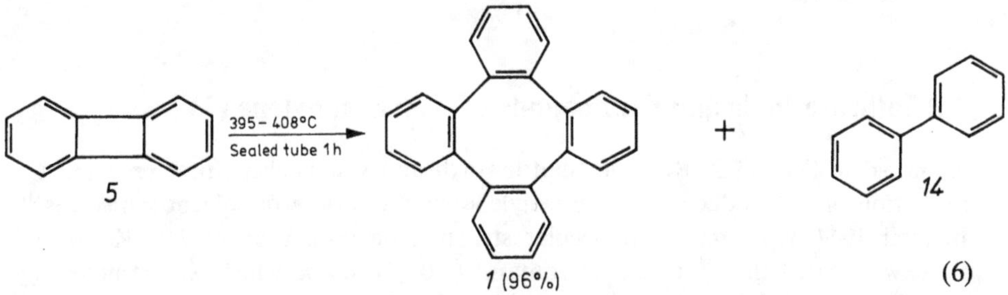

(6)

Subsequent study disclosed that pyrolysis of neat *5* in a sealed tube at 395—408 °C for 1 hour gave 96% yield of *1* and 4% of biphenyl *14* (Eq. 6) [15,16]. Pyrolysis of *5* at higher temperatures (430—445 °C) and for extended reaction times only resulted in thermal destruction of the initial product and thus a lower yields of *1* [15].

In view of the fact that almost all the other available methods only produce tetra-

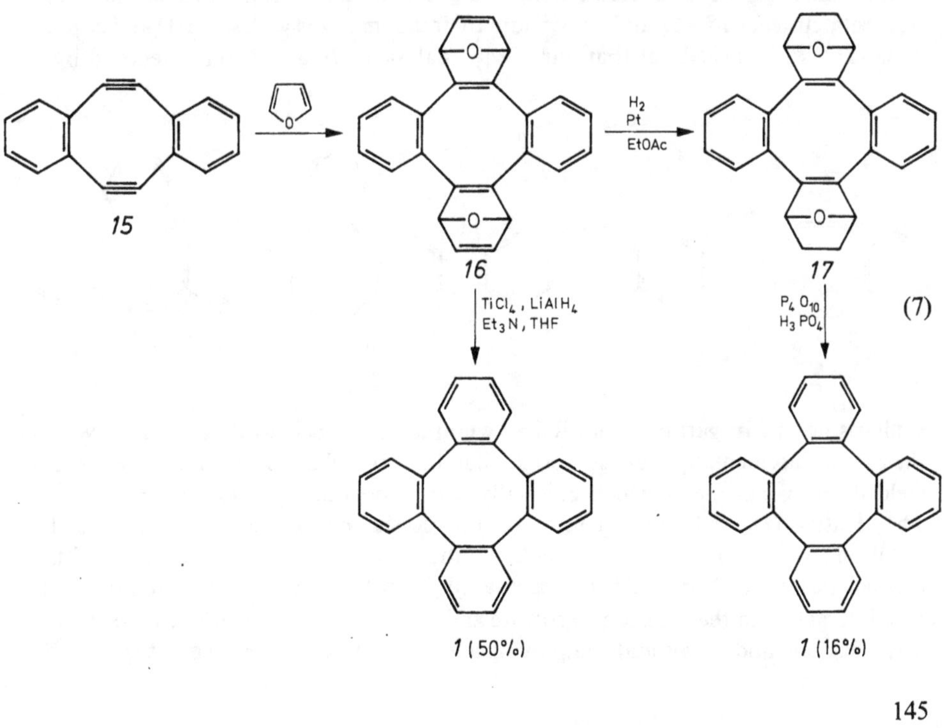

(7)

phenylene *1* in poor yields through complicated separation procedures, this pyrolytic synthesis can be regarded as the most practical method for the preparation of *1* in bulk quantities. More recently, it has been found that efficient pyrolysis of *5* to give *1* can be achieved at 250 °C with 2% of di-μ-chloro-bis(norbornadienerhodium) as a catalyst [17].

The diacetylene *15* readily underwent Diels-Alder cycloaddition with furan to furnish the endoxide *16*, which was hydrogenated catalytically to the tetrahydro compound *17*. Dehydration of *17* with polyphosphoric acid gave *1* in 16% yield (Eq. 7) [18]. Alternatively, deoxygenation of *16* with low valent titanium generated by reducing titanium tetrachloride with lithium aluminium hydride also afforded *1* in 50% yield [19].

4 Clathrate Inclusion Compounds of Tetraphenylene (1)

As noted in Section 2, Rapson, Shuttleworth and van Niekerk first reported the formation of 2:1 adducts of tetraphenylene with a variety of solvent molecules [3]. In their 1944 paper on the molecular structure determination of *1* [4], Karle and Brockway stated that 'the sample contained 10% benzene which was removed by placing the sample in the nozzle and heating it to 200° in a vacuum before electron diffraction photographs were taken.' Understandably, the nature of molecular association in these adducts drew no attention at that time, since the concept of clathration was not clarified until Powell's pioneering study of the hydroquinone (quinol) clathrates in 1947 [20]. Nevertheless, it is surprising that the problem lay dormant for the next several decades.

In the early 1980's we started our structural studies on the series of benzannelated cyclooctatetraenes *18*—*21* and *1* [21,22] and their π-complexes with silver(I) and copper (I) salts [23,24]. It turned out that our X-ray analysis of *19* and *1* were preceded by a

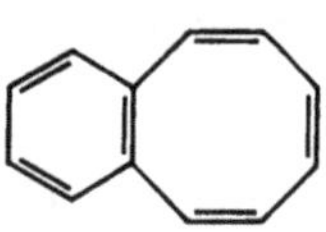
18

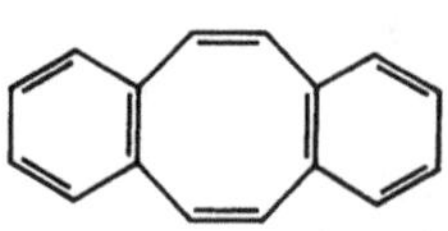
19

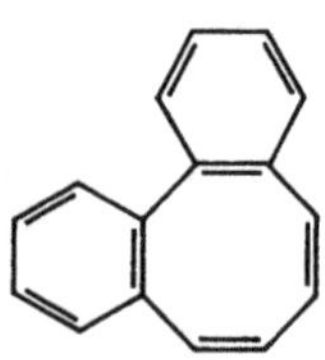
20

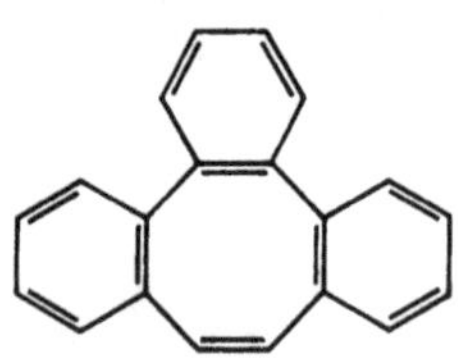
21

similar study by Irngartinger and Reibel who published their work in 1981 [7]. Nevertheless, we accidentally 're-discovered' that the crystallization of *1* in chloroform yielded tetragonal plates which gradually turned opaque upon exposure to air [25]. The density (D_m = 1.320 g cm^{-3}) of the crystalline product as measured by flotation in aqueous potassium iodide solution considerably exceeded that of the monoclinic form of *1* crystallized from ethanol. These findings were indicative of guest-host interaction in the chloroform solvate and prompted us to undertake a systematic crystallization and X-ray study employing a wide variety of potential guest species [25].

4.1 Crystal Data and Structural Characteristics

Clathrate crystals obtained by slow evaporation of a solution of *1* in suitable 'guest solvents' conform to the general formula $2\,C_{24}H_{16} \cdot G$, where G is a guest species ranging in size from methylene chloride to cyclohexane (Table 1). The tetraphenylene

Table 1. Crystal data for clathrate inclusion compounds of tetraphenylene, $2\,C_{24}H_{16} \cdot G$

			Volume (Å³)		
Guest solvent, G	*a* (Å)	*c* (Å)	Unit cell	Cavity	Guest
CH_2Cl_2	9.892(5)	18.46(1)	1806	77.7	82.0 (−120 °C)
Acetone	9.902(2)	18.491(6)	1813.0	81.2	
Tetrahydrofuran	9.906(1)	18.503(5)	1815.7	82.6	
CH_2Br_2	9.935(2)	18.546(6)	1830.6	90.0	95.1 (−90 °C)
$CHCl_3$	9.925(2)	18.593(3)	1831.5	90.5	103.8 (−88 °C)
Dioxan	9.968(1)	18.553(5)	1843.5	96.5	
2-Brompropane	9.973(1)	18.633(5)	1853.3	101.4	
1-Bromopropane	10.004(1)	18.647(4)	1866.2	107.8	
CCl_4	9.930(2)	18.932(6)	1866.8	108.2	116.6 (10 Kbar)
Benzene	10.069(1)	18.431(5)	1868.6	109.0	118.5 (−135 °C)
Cyclohexane	10.073(1)	18.712(2)	1898.6	124.1	140.3 (−158 °C)

clathrates, which easily lose their guest components in air, constitute an isomorphous series belonging to space group $P4_2/n$ (No. 86) with $Z = 2$ [25], and are morphologically distinguishable from pure host *1* which crystallizes from ethanol, methyl iodide, acetonitrile, carbon disulfide, diethyl ether, *n*-butanol, methyl ethyl ketone, acetic acid, ethyl acetate, nitrobenzene, and toluene in space group $C2/c$ (No 15) with $a = 15.628$, $b = 13.126$, $c = 16.369$ Å, $\beta = 100.56°$, $V = 3301.0$ Å³, $Z = 8$ and $D_x = 1.225$ g cm^{-3} [7]. To facilitate comparison with the clathrates, the *C*-centered cell of *1* may be transformed to a primitive unit cell of half its original size by the transformation:

$$a' = {}^1/_2(a + b)$$
$$b' = {}^1/_2(a - b)$$
$$c' = -c$$

For this reduced unit cell, $a' = b' = 10.205$, $c' = 16.369$ Å, $\alpha = \beta = 81.93$, $\gamma = 80.05°$, $V = 1650.5$ Å³ and $Z = 4$. Crystal data for the clathrate inclusion compounds of *1* are tabulated in Table 1 in the order of increasing unit-cell volumes [25].

Two stereoviews of the molecular packing in a representative clathrate, namely $2\,C_{24}H_{16} \cdot CCl_4$ [26], are shown in Figs. 1 and 2. With the unit-cell origin at $\bar{1}$, the host and guest molecules occupy Wyckoff positions 4(*e*) of point symmetry 2 and 2(*a*) of point symmetry $\bar{4}$, respectively. Eight host molecules cluster around a spheroidal cavity centered at $({}^1/_4, {}^1/_4, {}^1/_4)$, which accommodates the guest molecule. The latter necessarily satisfies the $\bar{4}$ site symmetry, generally through orientational disorder. All intermolecular interactions are of the van der Waals type.

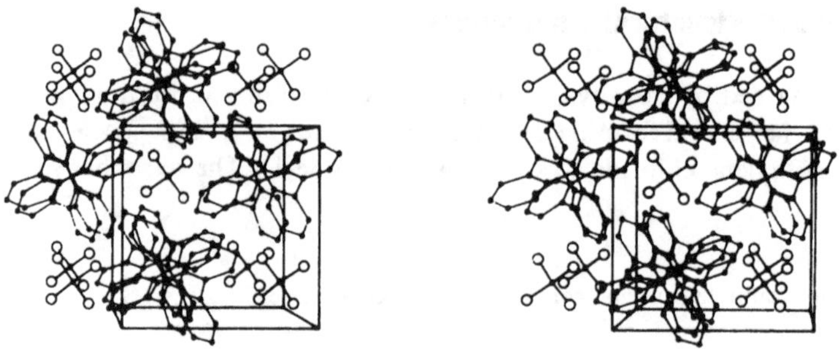

Fig. 1. Stereoview of the molecular packing in the 2 $C_{24}H_{16} \cdot CCl_4$ clathrate illustrating the environment of a guest species located at ($^1/_4$, $^1/_4$, $^1/_4$). The disordered CCl_4 molecule is shown in its preferred orientation, and H atoms have been omitted for clarity. The unit-cell origin lies at the upper left corner, with **a** pointing from left to right, **b** downwards, and **c** away from the reader

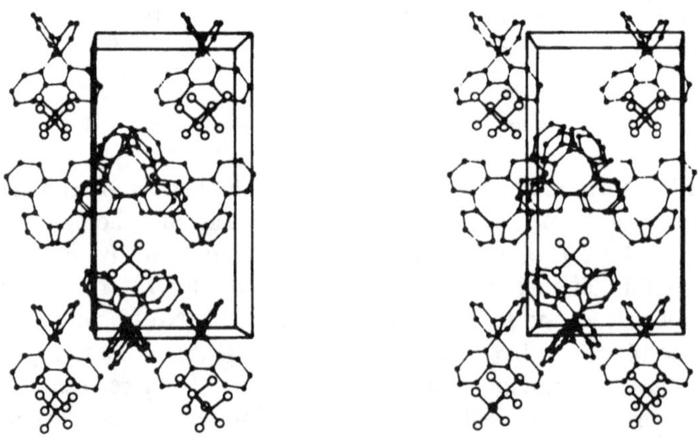

Fig. 2. The molecular packing of 2 $C_{24}H_{16} \cdot CCl_4$ viewed in a second axial direction. The origin of the unit cell lies at the lower left corner, with **a** pointing towards the reader, **b** from left to right, and **c** upwards

4.2 The Host Molecule

A consistent system of numbering is used for both crystalline tetraphenylene [7] and its chloroform [25], carbon tetrachloride [26], benzene [27], and cyclohexane [27] inclusion complexes (Fig. 3). A crystallographic C_2 axis passes through the centers of a pair of single bonds, C(1)—C(1′) and C(12)—C(12′), of the eight-membered ring.

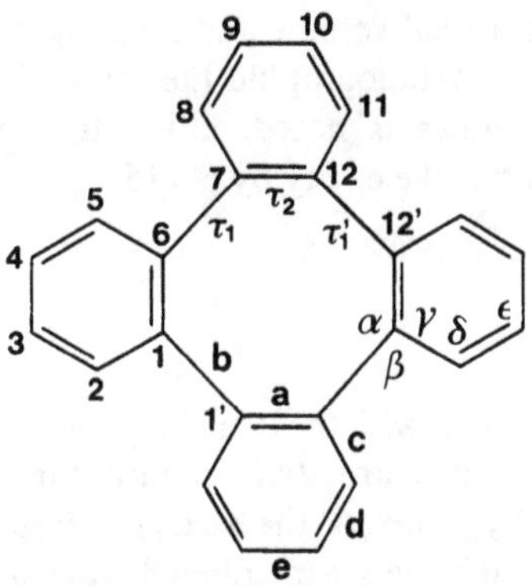

Fig. 3. Structural formula of tetraphenylene *1* showing the numbering system used and labelling of the chemically equivalent bond lengths, bond angles, and torsion angles. Atoms labelled with primes are related to the reference atoms by two-fold rotational symmetry

Chemically equivalent bond lengths, bond angles, and selected torsion angles are averaged in Table 2; apart from the larger difference between τ_1 and τ_1' values in the clathrates, there is good agreement throughout this series of related compounds. In all instances the bond lengths and bond angles follow the trends: $b \gg a \sim c \sim d > e$, $\alpha \gg \beta \sim \gamma$, and $\delta > \varepsilon > \gamma$.

Table 2. Comparison of averaged bond lengths, bond angles, and selected torison angles of the tetraphenylene molecule in the host compound and the clathrates

Type[a]	$C_{24}H_{16}$	$2\,C_{24}H_{16}\cdot$ benzene	$2\,C_{24}H_{16}\cdot$ cyclohexane	$2\,C_{24}H_{16}\cdot CHCl_3$	$2\,C_{24}H_{16}\cdot CCl_4$
a	1.400(3)	1.41(1)	1.399(7)	1.40(1)	1.396(6)
b	1.494(3)	1.48(1)	1.491(10)	1.50(1)	1.499(8)
c	1.397(3)	1.39(1)	1.389(9)	1.41(1)	1.401(8)
d	1.381(4)	1.37(1)	1.380(9)	1.39(1)	1.383(8)
e	1.373(4)	1.35(1)	1.372(9)	1.37(1)	1.367(7)
α	122.5(2)	122.3(8)	122.6(5)	123.1(6)	123.0(4)
β	118.6(2)	119.6(9)	118.8(5)	117.8(6)	118.0(4)
γ	118.8(2)	117.9(9)	118.5(6)	119.0(7)	118.9(4)
δ	121.3(2)	122.6(9)	121.8(6)	121.0(7)	121.2(4)
ε	119.9(2)	119.4(9)	119.7(6)	120.0(8)	120.2(6)
τ_1	64.7(3)	65(1)	62.6(7)	60(1)	60.2(5)
τ_1'	−67.6(3)	−68(1)	−69.2(7)	−69(1)	−69.8(5)
$\lvert\tau_2\rvert$	1.0(3)	2(1)	2.3(9)	2(1)	3.6(5)

[a] Types as defined in Fig. 3. τ_1 and τ_2 denote torsion angles about single and double bonds, respectively, in the eight-membered ring; for the clathrates, τ_1' is the average of the two torsion angles about C(1)—C(1′) and C(12)—C(12′).

4.3 The Clathration Cavity

This takes the general shape of an oblate spheroid with its minor axis parallel to the *c* axis, the limiting shape being a sphere of free diameter 7.2—7.4 Å in the carbon tetrachloride clathrate [26]. The cavity size undergoes significant changes as the host cagework adapts itself to the steric requirements of the various guests. Table 1 also presents a comparison of the volume of the cavity (calculated by subtracting the volume of the primitive cell of tetraphenylene from the clathrate cell volume and dividing by two) and the volume of the guest as reported in the crystallographic literature [27]. An excellent correlation between the two sets of values is noted, although the estimated volume of the guest consistently exceeds that of the cavity by 6—15%.

4.4 The Guest Molecules

The sizes and shapes of the guest molecules can be correlated with the two independent crystallographic axes. In Table 3, the various clathrates are divided into three groups according to the nature of the guest species [27]. In group (A) the *a* axis increases from the five-membered ring of tetrahydrofuran, through the six-membered rings of dioxan and benzene, to the chair structure of cyclohexane. The same variation in the *c* axis is found for the non-planar guests, but a significant contraction in *c* occurs in the case of the planar guest benzene, which has its principal molecular axis parallel to *c*.

Several general patterns may be noted in group (B). The CH_2Cl_2 and CH_3COCH_3 clathrates have similar lattice dimensions because the size of the methyl group is close to that of chlorine, and the two hydrogen atoms in the former guest species occupy roughly the same space as the oxygen atom in the latter. In a similar way, the two bromine atoms in CH_2Br_2 have roughly the same spatial requirement as the three chlorine atoms in $CHCl_3$, so the two clathrates have nearly the same unit-cell volume and differ little in their lattice dimensions. The measured crystal data clearly show that the 2-bromopropane guest is intermediate in size between $CHCl_3$ and CCl_4;

Table 3. Crystal data ordered according to groups of guest molecules

Group	Guest	*a* (Å)	*c* (Å)
(A)	Tetrahydrofuran	9.906(1)	18.503(5)
	Dioxan	9.968(1)	18.553(5)
	Benzene	10.069(1)	18.431(5)
	Cyclohexane	10.073(1)	18.712(2)
(B)	CH_2Cl_2	9.892(5)	18.46(1)
	CH_3COCH_3	9.902(2)	18.491(6)
	CH_2Br_2	9.935(2)	18.546(6)
	$CHCl_3$	9.925(2)	18.593(3)
	2-Bromopropane	9.973(1)	18.633(5)
	CCl_4	9.930(2)	18.932(6)
(C)	1-Bromopropane	10.004(1)	18.647(4)

the bulk of the bromine atom is reflected mainly in an increase in the *a* axis, whereas the longest *c* axis is observed in the CCl_4 clathrate.

The linear-chain molecule 1-bromopropane which alone comprises group (C), as compared to all guest species in group (B) expect CCl_4, corresponds to lengthening of both axes, the increase in *a* being more significant.

Taken as a whole, the measured crystallographic data are consistent with the conclusion that a given guest species tends to spread its bulk in a plane parallel to (001), and the tetraphenylene cagework readily adapts itself to accommodate the preferred orientation of the guest molecule in the cavity.

As noted previously, the guest species in a tetraphenylene clathrate must satisfy the required $\bar{4}$ site symmetry. In the chloroform clathrate, the electron density maxima corresponding to the enclosed $CHCl_3$ molecule could not be interpreted in terms of an easily visualized model [25]. In the benzene clathrate, the guest molecule lies parallel to (001) with two equally probable orientations mutually rotated by 30° about its principal symmetry axis [27]. For the cyclohexane clathrate, the Fourier peaks about ($^1/_4$, $^1/_4$, $^1/_4$) could be fitted to four *chair* cyclohexanes (two pairs, each having the four in-plane carbon atoms in common) [27]. It is notable that in the carbon tetrachloride clathrate, the encaged CCl_4 molecule also exhibits orientational disorder, despite the fact that its size and shape ideally match the specifications for an ordered guest species [26]. In the model adopted for refinement, the guest molecule has a preferred orientation of 50% population, with the residual scattering accounted for by the superposition of two or more minor orientations.

4.5 General Conclusions

From a geometrical point of view, the clathration cavity may be visualized as created by an expansion of the reduced unit cell of the host compound (see Sect. 4.1) in conformity with the imposed tetragonal symmetry. Interestingly, this involves a considerable lengthening of the *c'* axis accompanied by a slight contraction of the other two axes, the interaxial angles increasing by 8—10° to become right angles. The measured lattice dimensions and X-ray crystal structures of the tetraphenylene clathrates have established that the host cagework is capable of adapting itself to the subtle steric requirements of the enclosed guest molecules. The significant difference in τ_1 and τ_1' for the host molecule in the clathrates is consistent with the recongnition that twofold molecular symmetry plays a dominant role in the architecture of many clathration lattices consolidated by van der Waals attraction and/or hydrogen bonding [28,29].

5 Synthesis of Substituted Tetraphenylenes

Two approaches have been devised to prepare tetraphenylene derivatives. The first approach involves electrophilic aromatic substitution which can be performed on *1* because its benzene rings are not in conjugation with each other and therefore behave as rather normal benzene nuclei. The drawback of this methodology is that it is difficult to predict the position of substitution. Alternatively, the substituentes can be appropriately positioned on the starting materials before the actual construction of the tetraphenylene skeleton. This approach usually leads to tetraphenylene derivatives with substituents on desired positions. The following examples suffice to show how these two approaches were executed.

Tetraphenylene *1* was shown to undergo electrophilic aromatic substitution with bromine to provide bromotetraphenylene *22* (Eq. 8) [3].

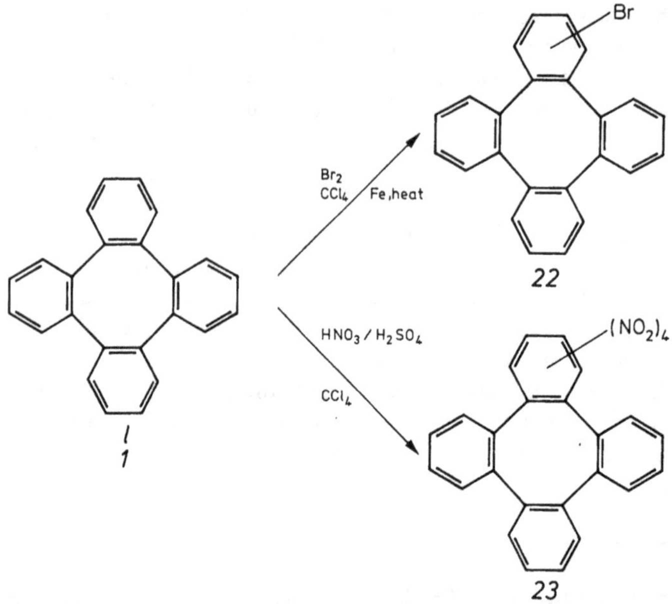

(8)

(9)

Moreover, nitration of *1* furnished tetranitrotetraphenylene *23* (Eq. 9). [3]. Upon trituration with excess carbon tetrachloride, *23* gave crystals of molecular formula $C_{24}H_{12}N_4O_8 \cdot CCl_4$ which suggests the possibility of inclusion complex formation [3]. However, the positions of the bromine atom of *22* and the nitro groups of *23* have not been ascertained.

An acylation reaction could also be performed on tetraphenylene *1*. Thus, the reaction of *1* with acetyl chloride gave 2-acetyltetraphenylene *24*, which was converted

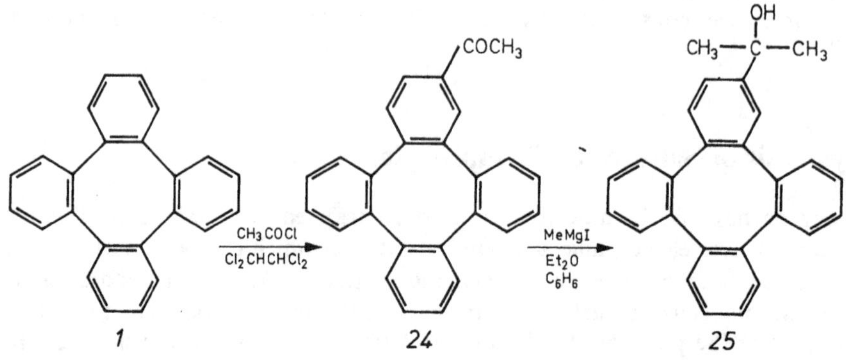

(10)

to the alcohol *25* by reaction with methyl magnesium iodide (Eq. 10)[30]. An extraordinarily low barrier to ring-inversion (5.7 ± 1 Kcal/M) was found for *25* from variable temperature NMR spectroscopy[30].

2-Iodo-3-nitrobiphenyl *26* could be converted to the iodonium iodide *27* in 75% yield *via* standard procedure. Upon heating with copper(I) oxide at 300—340 °C, 0.56% yield of 1,16-dinitrotetraphenylene *28* as well as 1-nitrobiphenylene *29* were isolated (Eq. 11)[31].

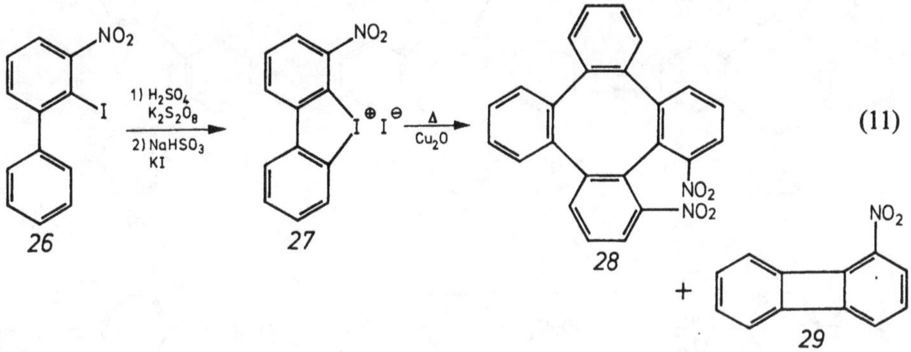

(11)

Similarly, iodides *30* were transformed to *31*, which yielded diiodide *32* on heating (Eq. 12). The dilithiobiphenyl *33a*, generated from *32a*, was quenched by copper(II) bromide to provide a mixture of *34a* and *35a*. On the other hand, *33b* gave separable products *34b* and *35b*[32]. (Table 4)

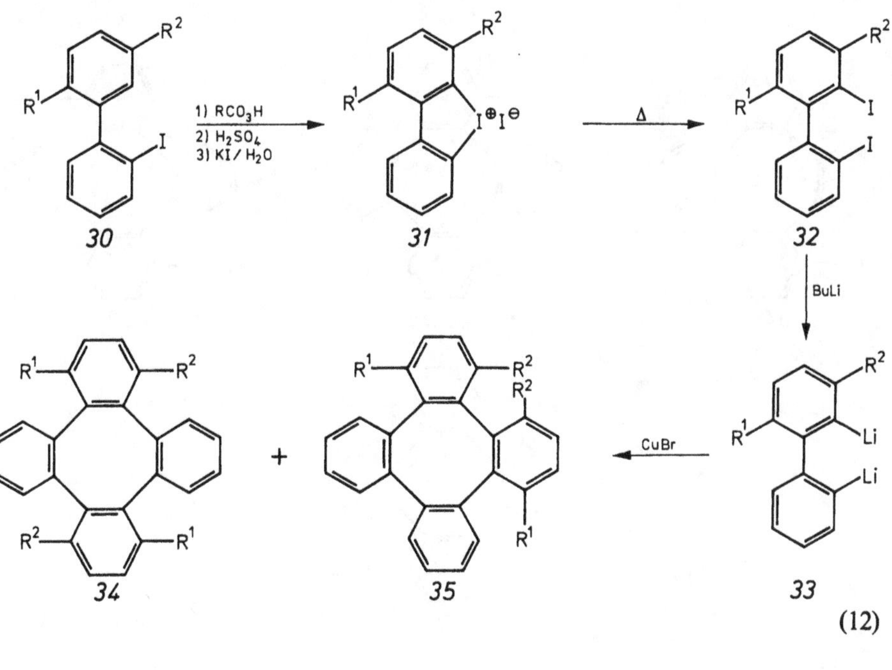

(12)

Oxidation of *34b* and *35b* with potassium permanganate afforded the tetraacids *36* and *38* in 76% and 80% yields, respectively (Eq. 13 and 14). The acids *36* and *38* were able to react with diazomethane to afford their corresponding esters *37* and *39* in quantitative yields [32]. On the other hand, the inseparable mixture of *34a* and *35a* could be converted to their acids *40* and *41* which were also difficult to separate (Eq. 15) [32].

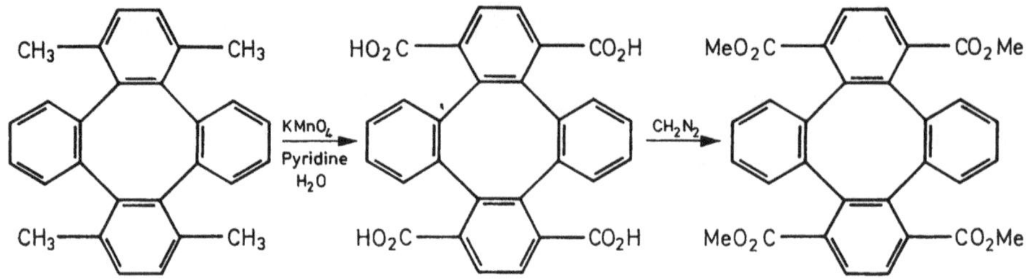

34 b *36* (76%) *37* (13)

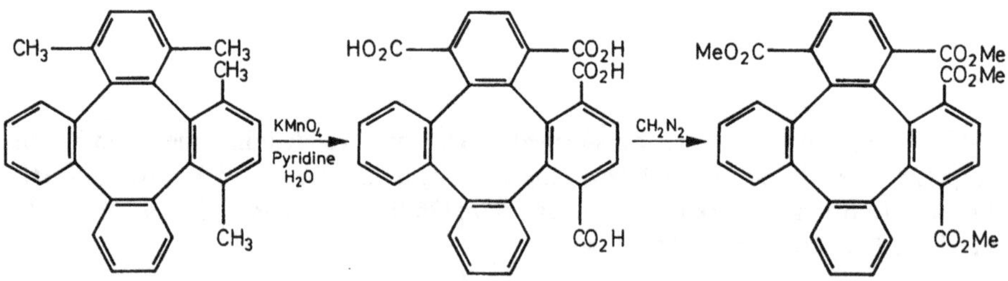

35 b *38* (80%) *39* (14)

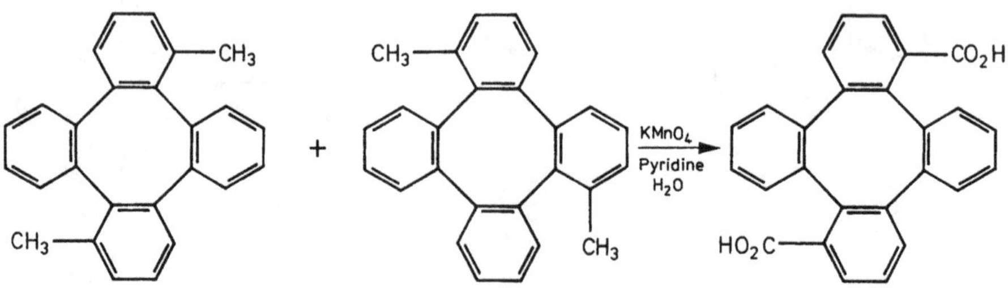

34 a *35 a* *40*

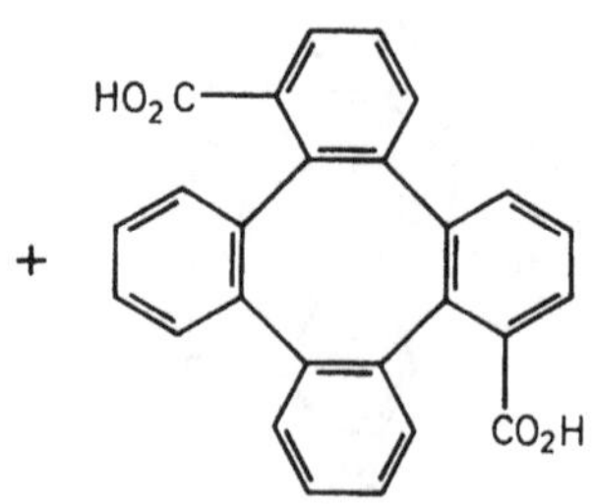

41 (15)

Table 4.

Compound	R¹	R²	Yield (%)
31a	Me	H	75
31b	Me	Me	83
32a	Me	H	65
32b	Me	Me	75
34a + *35a*	Me	H	20
34b	Me	Me	3
35b	Me	Me	3

6 Synthesis of Tetraphenylenes Fused with Carbocycles and Heterocycles

Diacetylene *15* [18] readily reacted with isobenzofuran *42* [33] to furnish the endoxide *43*, which was deoxygenated by low valent titanium to dibenzofused tetraphenylene

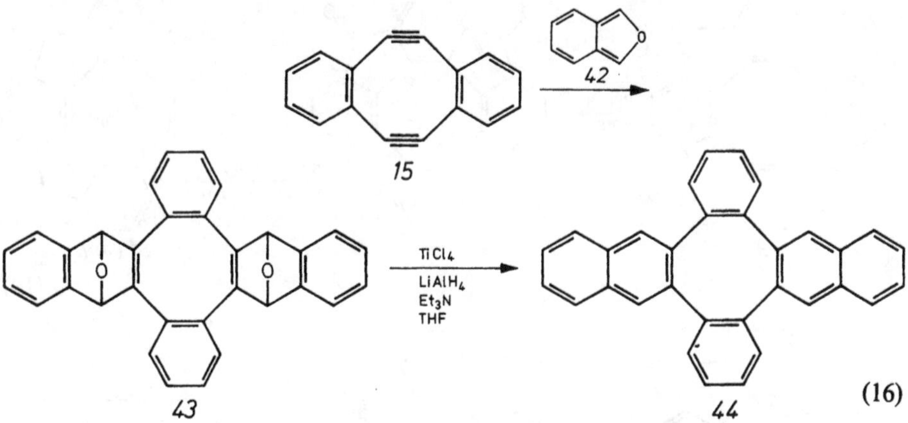

(16)

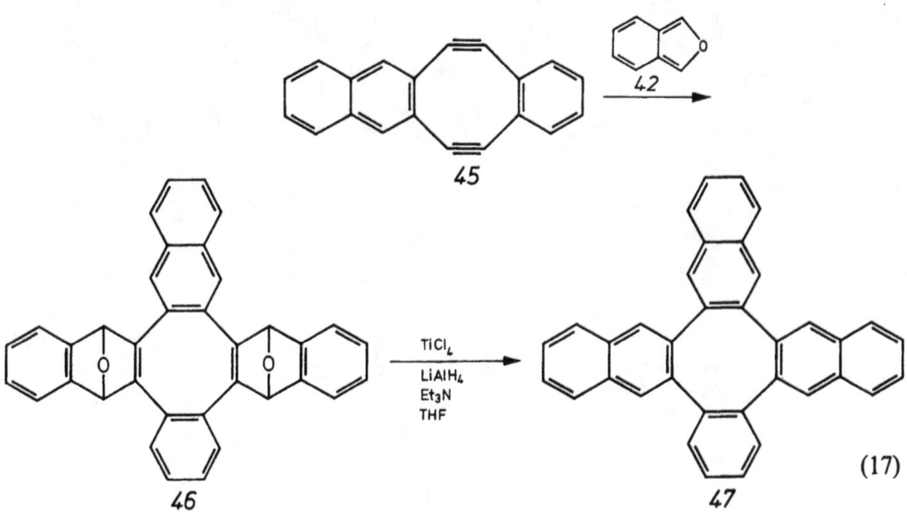

(17)

44 (Eq. 16) [34]. Similarly, *45* could be converted to the tribenzofused tetraphenylene *47 via* the endoxide *46* (Eq. 17) [34]. Both *44* and *47* are extremely stable crystalline compounds.

Intramolecular Friedel-Crafts acylation of *37* yielded quantitively the tetraketone *48*, which could be reduced under Huang Minlon condition to 3,6,9,12-tetrahydro-

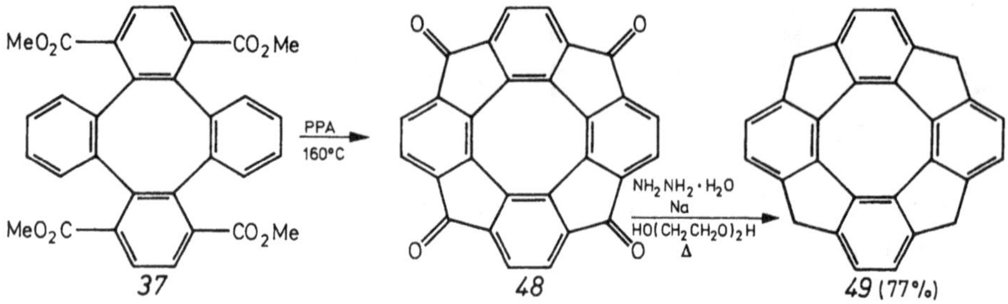

(18)

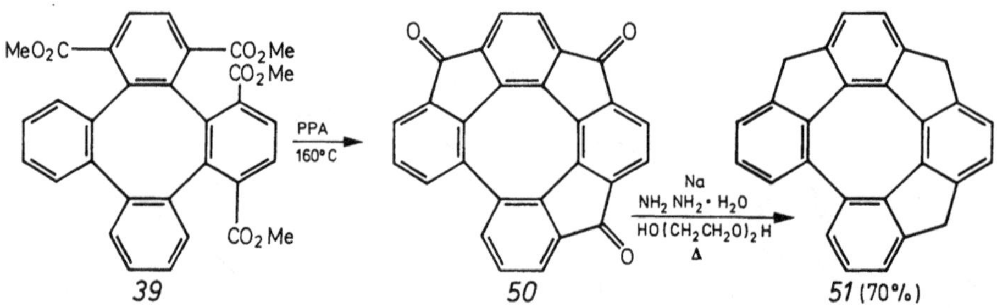

(19)

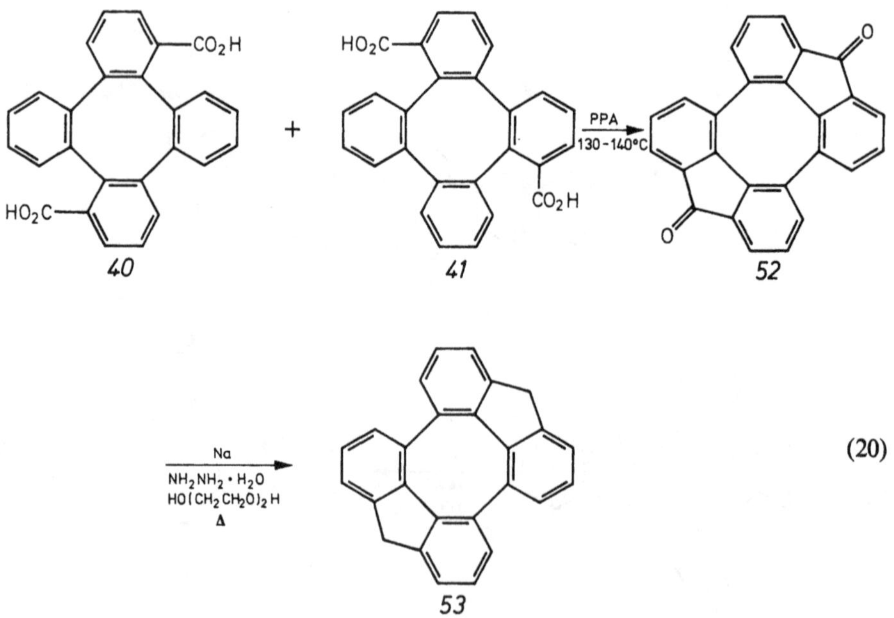

(20)

tetracyclopenta[*def, jkl, pqr, vwx*]tetraphenylene *49* in 77% yield (Eq. 18) [32,35]. It is important to note that both showed increased intensity as well as bathochromic shift in their UV spectra. This phenomenon is attributable essentially to the steric constraints exerted by the fused five-membered rings which could force *48* and *49* towards planarity.

Similarity, *39* was converted in a total yield of 70% to *51 via 50*. Both *50* and *51* were believed to be essentially planar [32,35]. The inseparable mixture of acids *40* and *41* was treated with polyphosphoric acid to give 37% of the diketone *52*, which was then reduced to *53* in 72% yield [32]. Interestingly, *53* was proved to be non-planar by proton NMR [36].

Bis-Wittig reaction between *54* and *55* afforded a low yield of *56* which cyclized to dibenzo[*def, pqr*]tetraphenylene *57* in 65% yield by oxidative photolysis procedure (Eq. 21a) [37]. Alternatively, operating under high dilution condition gave the disulfide macrocycle 60 from *58* and *59* (Eq. 21b). Standard procedures (sulfonium salt formation, Stevens rearrangement, sulfoxide formation, elimination) also furnished *56* in 33% total yield. Oxidative photocyclization again allowed the conversion of *56* to *57* in 63% yield [38]. An X-ray crystallographic study revealed that the central eight-membered ring of *57* is in a distorted tub form [39].

Polymerization of α-naphthoquinone *61* induced by Lewis acid gave an oxygen-bridged tetramer *58* in 95% yield (Eq. 22) [40,41]. The UV spectrum of *62* showed it to be a planar molecule [40,41], although further confirmation is necessary. Similarly, substituted *p*-benzoquinones *63* tetramerized to *64* (Eq. 23) (Table 5) [42,43]. The

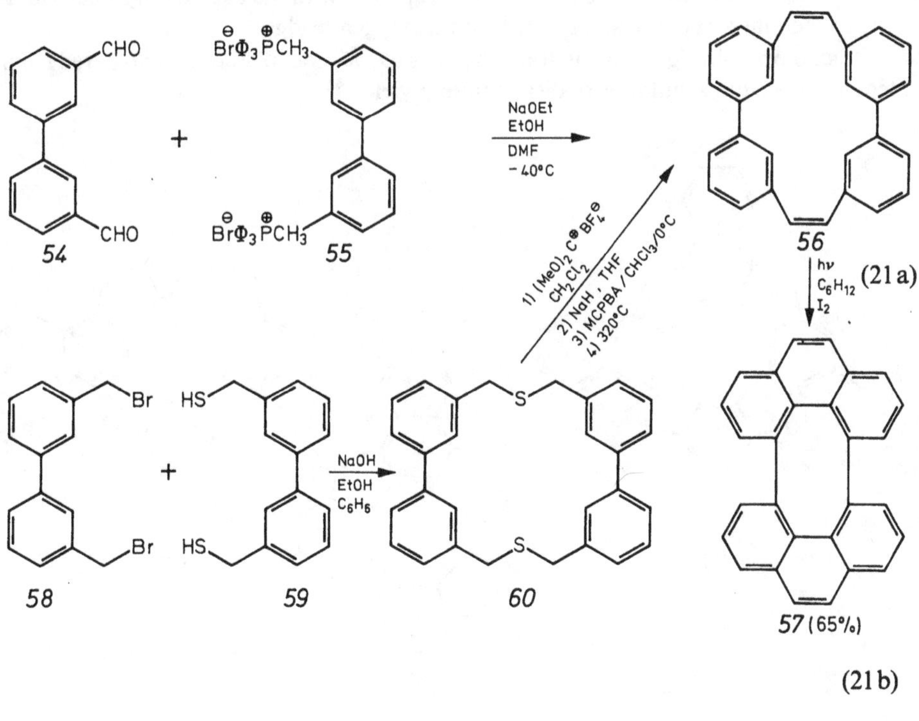

(21a)

(21b)

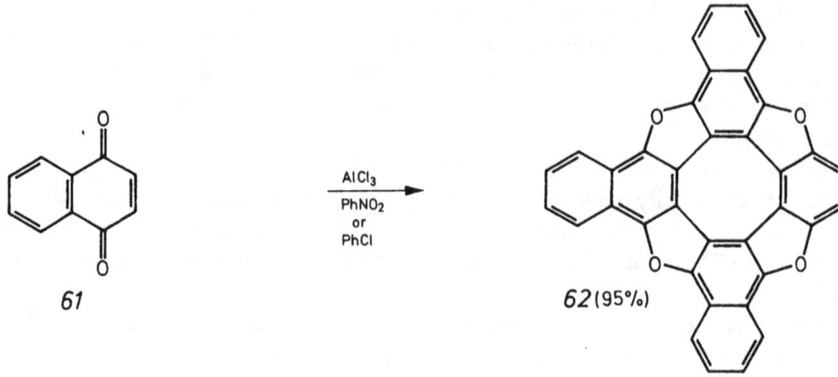

61 62 (95%)

(22)

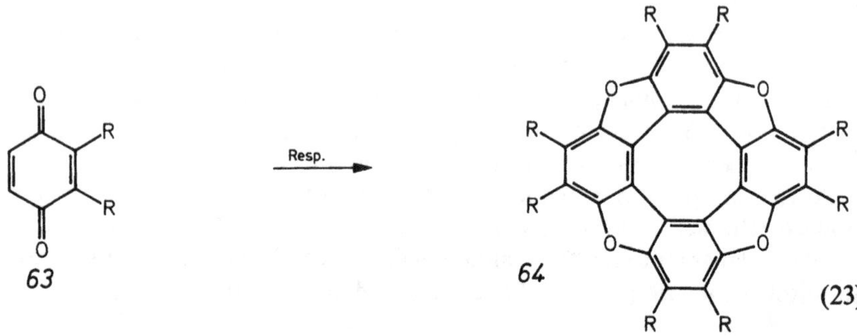

63 64

(23)

central eight-membered ring should also be coplanar with the bezene rings due primarily to the inherent constraining effect of the oxygen-bridges.

Other oxygen-bridged benzotetraphenylenes *65*, *66*, and *67* were also prepared from tedious procedures and in very disappointing yields [41].

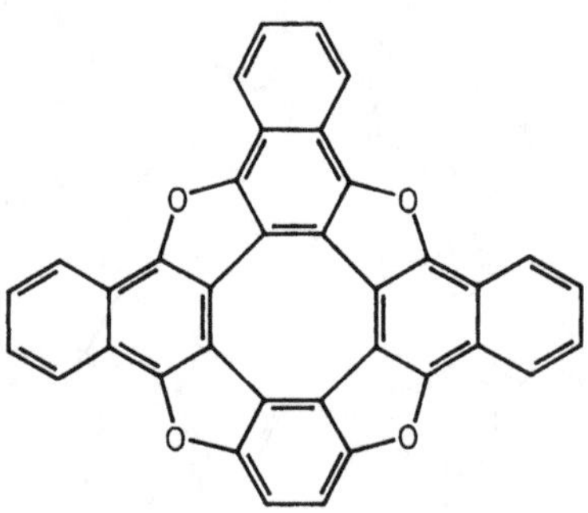

65

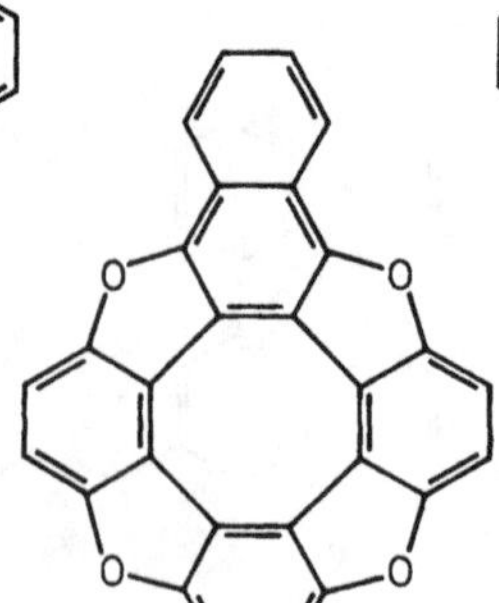

66

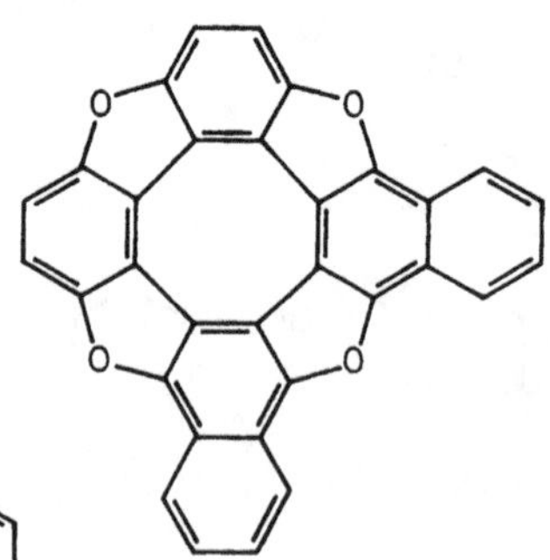

67

Table 5.

63 to *64*	R	R	Reaction condition	Yield (%)
a	Me	Me	$AlCl_3/PhNO_2$	60
a	Me	Me	$H_2SO_4/HOAc/H_2O$	30
b	Pr^n	Pr^n	$AlCl_3/PhNO_2$	60
c	$-(CH_2)_4-$		$AlCl_3/PhNO_2$	39

7 Synthesis of Heterocyclic Counterparts of Tetraphenylene

An excellent account reviewing the heterocyclic counterparts of tetraphenylene *1* has appeared [44].

The dibromides *68* and *69* could be lithiated to their corresponding dilithiothiophenes *70* and *71*, which reacted with copper(II) chloride to provide the tetrathiophene *72* in 25% and 18% yields, respectively (Eq. 24). The tetrathiophene *72* was

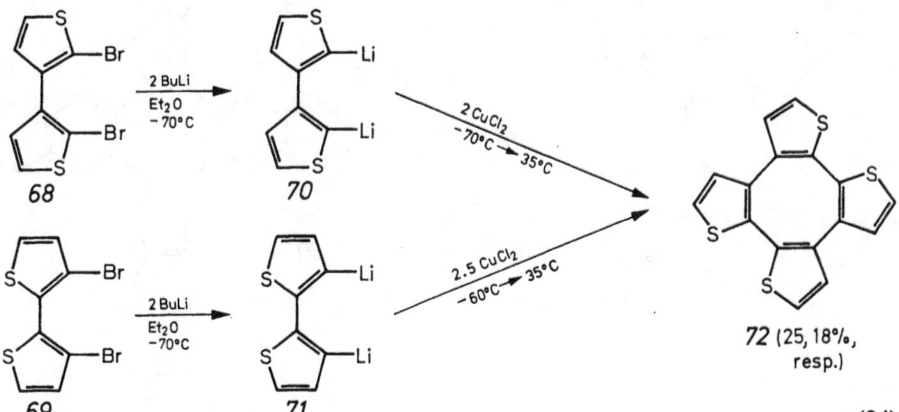

(24)

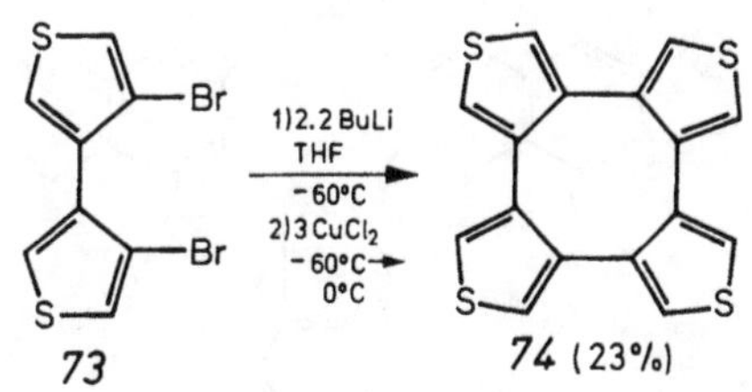

(25)

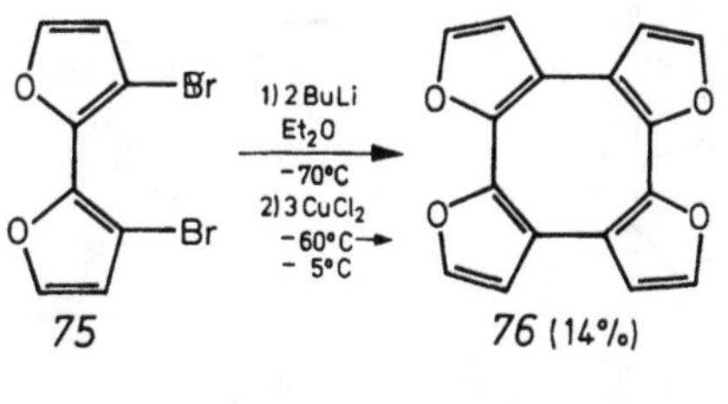

(26)

proposed to be a non-planar molecule on the basis of spectroscopic studies [45,46].

The other tetrathiophene isomer *74* was prepared accordingly from *73* which resulted in 23% yield (Eq. 25) [46,47]. Similarily, the conversion of *75* to the tetrafuran *76* was achieved in 14% yield (Eq. 26) [46,47]. Moreover, 5,5'-bipyrimidinyl *77* could be cyclodimerized to *78* in 12% yield upon reaction with excess lithium diisopropylamide (Eq. 27) [46,47].

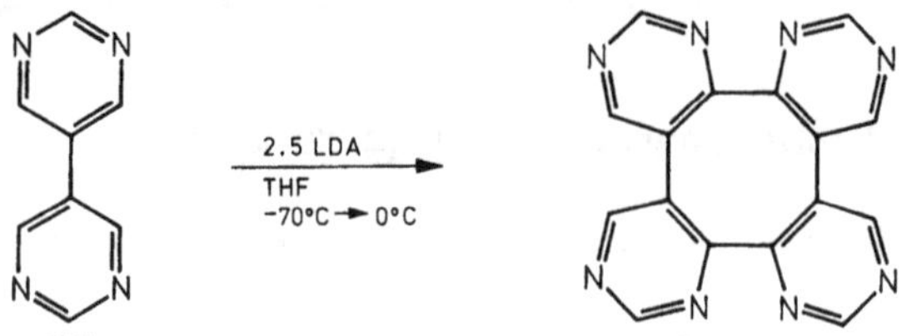

77 *78* (12%) (27)

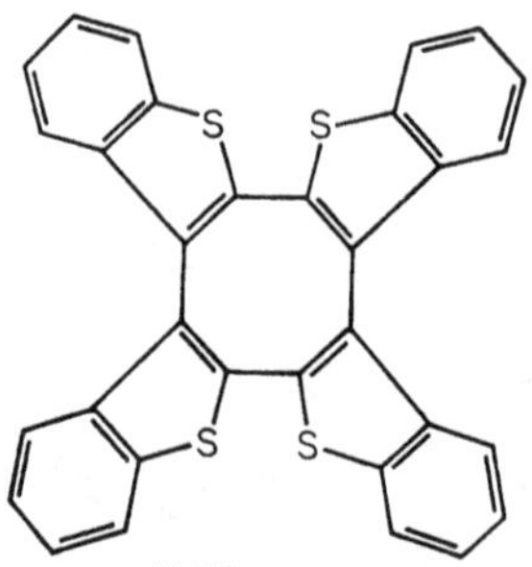

79 [46,47]

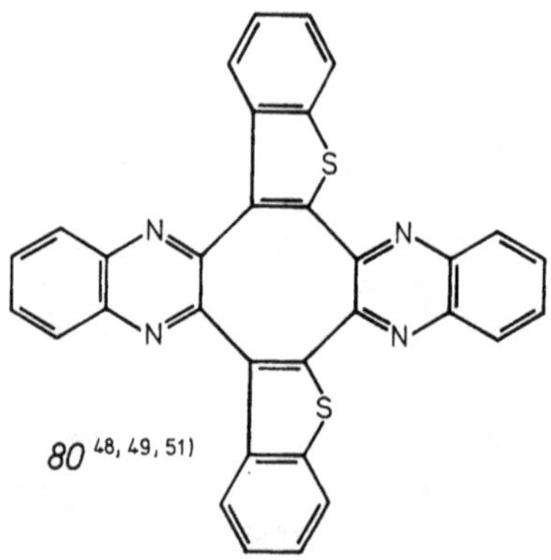

80 [48,49,51]

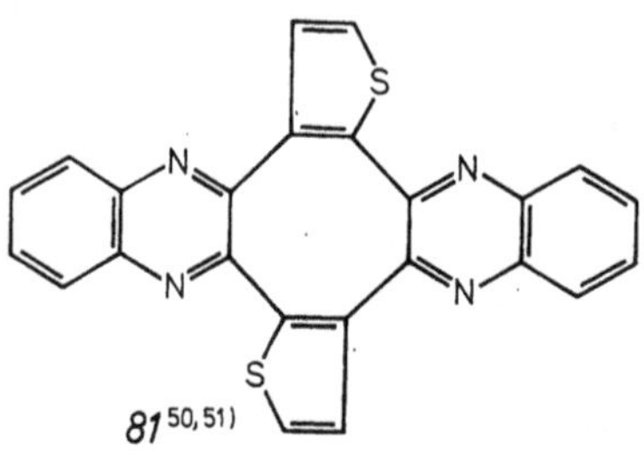

81 [50,51]

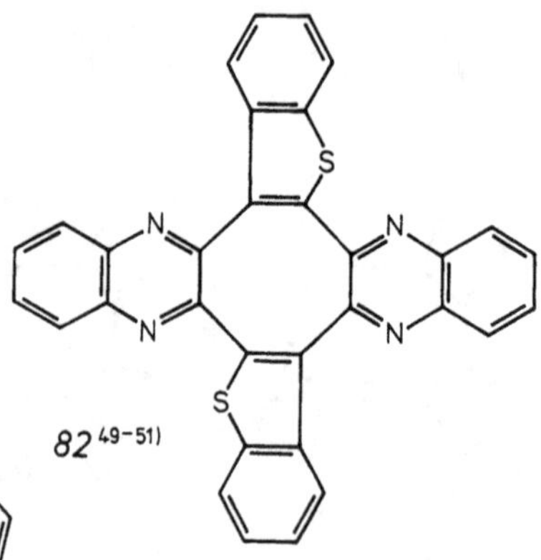

82 [49–51]

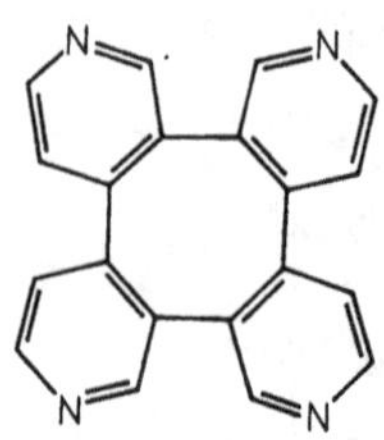

83 [44]

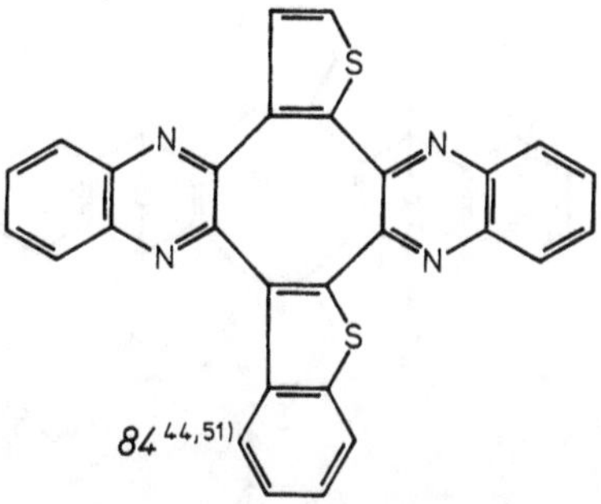

84[44,51]

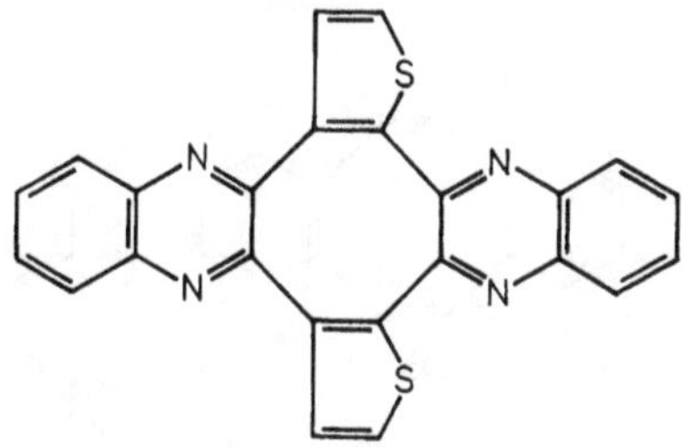

85[44]

Other heterocyclotetraaromatic compounds related to *1*, such as *79*—*85*, have been synthesized by the same general strategy.

8 Synthesis of Endoxide Derivatives of Tetraphenylene

The oxygen-bridged tetrahydrotetraphenylene derivatives *16*[18,19], *43*[3,4] and *46*[34] are potential host molecules in clathrate formation.

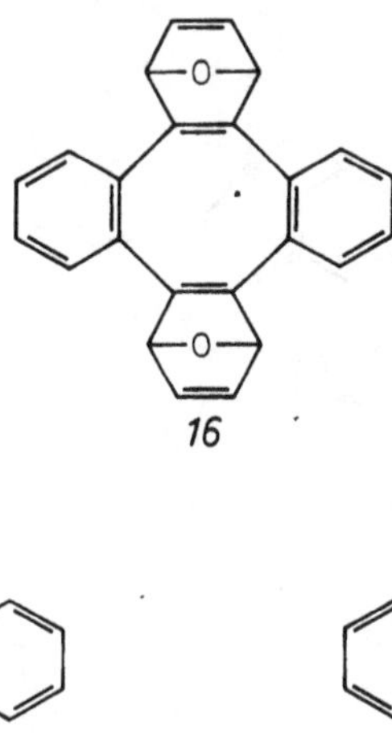

16

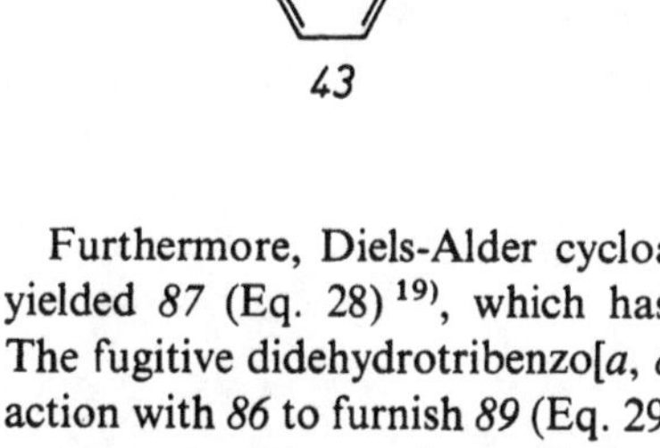

43

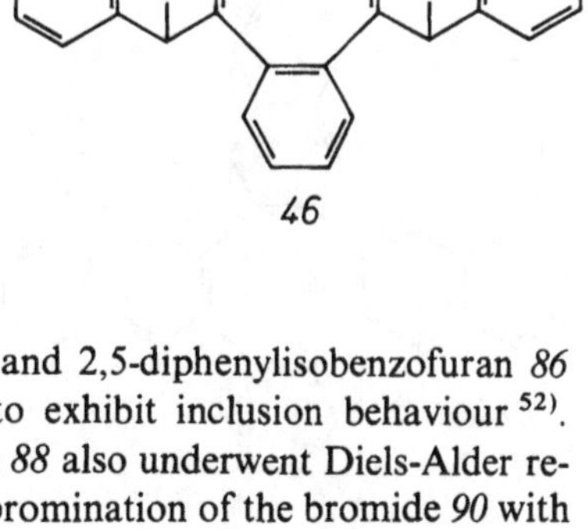

46

Furthermore, Diels-Alder cycloaddition of *15* and 2,5-diphenylisobenzofuran *86* yielded *87* (Eq. 28)[19], which has been found to exhibit inclusion behaviour[52]. The fugitive didehydrotribenzo[*a*, *c*, *e*]cyclooctene *88* also underwent Diels-Alder reaction with *86* to furnish *89* (Eq. 29)[53]. Dehydrobromination of the bromide *90* with excess potassium *tert*-butoxide in the presence of *86* afforded the trapping product *91* (Eq. 30)[54].

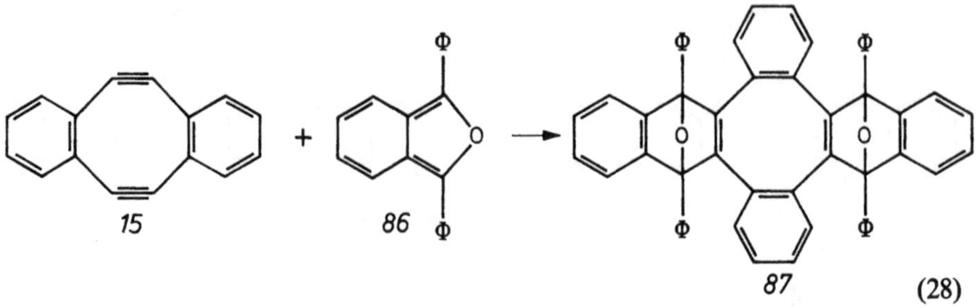

(28)

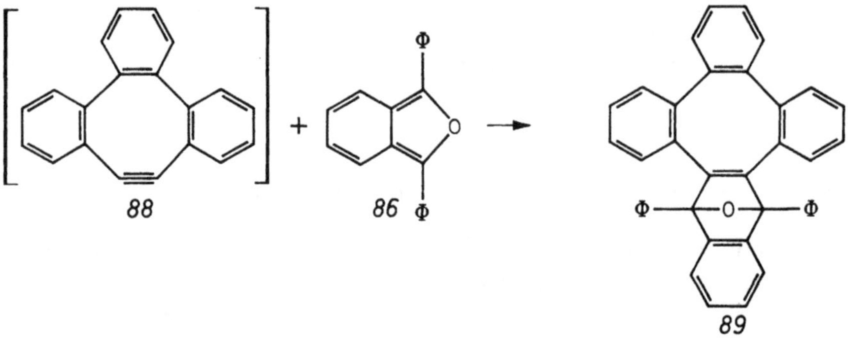

(29)

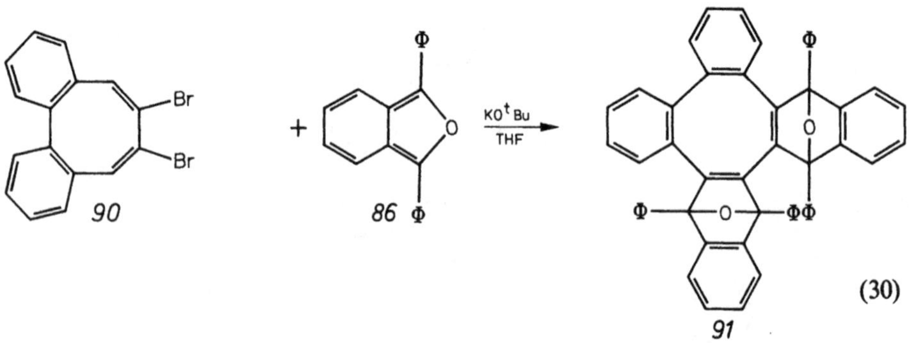

(30)

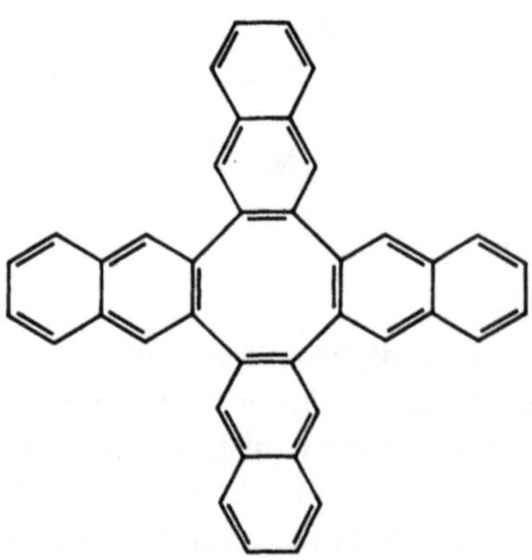

92

9 Concluding Remarks

Systematic X-ray crystallographic studies have provided a structural rationale for the inclusion behaviour of tetraphenylene, leading to a semi-quantitative assessment of the effective molecular sizes and shapes of a wide range of encaged guest species. However, much work remains to be done in the spectroscopic, thermodynamic and theoretical investigation of this class of clathrates.

The related carbocycles *44*, *47* and *92* (yet to be synthesized), as well as the endoxides *16*, *43*, *46*, *89* and *91* are potentially useful host systems, and their inclusion properties are currently under investigation.

10 References

1. Sircar, A. C., Majumdar, J. N.: J. Indian Chem. Soc. *5*, 417 (1928)
2. Kuhn, R.: Liebigs Ann. Chem. *475*, 131 (1929)
3. Rapson, W. S., Shuttleworth, R. G., van Niekerk, J. N.: J. Chem. Soc. *1943*, 326.
4. Karle, I. L., Brockway, L. O.: J. Am. Chem. Soc. *66*, 1974 (1944)
5. Rapson, W. S., Schwartz, H. M., Stewart, E. T.: J. Chem. Soc. *1944*, 73
6. McEwen, K. L., Longuet-Higgins, H. C.: J. Chem. Phys. *24*, 771 (1956)
7. Irngartinger, H., Reibel, W. R. K.: Acta Crystallogr. *B37*, 1724 (1981)
8. Wittig, G., Bickelhaupt, F.: Chem. Ber. *91*, 883 (1958)
9. Winkler, H. J. S., Wittig, G.: J. Org. Chem. *28*, 1733 (1963)
10. Wittig, G., Lehmann, G.: Chem. Ber. *90*, 875 (1957)
11. Wittig, G., Klar, G.: Liebigs Ann. Chem. *704*, 91 (1967)
12. Wittig, G.: Quart. Rev. Chem. Soc. *20*, 191 (1966)
13. Cade, J. A., Pilbeam, A.: J. Chem. Soc. *1964*, 114
14. Chatt, J., Guy, R. G., Watson, H. R.: ibid. *1961*, 2332
15. Lindow, D. F., Friedman, L.: J. Am. Chem. Soc. *89*, 1271 (1967)
16. Friedman, L., Lindow, D. F.: ibid. *90*, 2324 (1968)
17. Puglisi, O., Bottino, F. A., Recca, A., Stille, J. K.: J. Chem. Research(S) *1980*, 216
18. Wong, H. N. C., Sondheimer, F.: Tetrahedron *37*(S1), 99 (1981)
19. Xing, Y. D., Huang, N. Z.: J. Org. Chem. *47*, 140 (1982)
20. Palin, D. E., Powell, H. M.: J. Chem. Soc. *1947*, 208
21. Li, W.-K., Chiu, S.-W., Mak, T. C. W., Huang, N. Z.: J. Mol. Structure (THEOCHEM) *94*, 285 (1983)
22. Huang, N. Z., Mak, T. C. W.: J. Mol. Structure *101*, 135 (1983)
23. Mak, T. C. W., Ho, W. C., Huang, N. Z.: J. Organometal. Chem. *251*, 413 (1983)
24. Mak, T. C. W., Wong, H. N. C., Sze, K. H., Book, L.: ibid. *225*, 123 (1983)
25. Huang, N. Z., Mak, T. C. W.: J. Chem. Soc., Chem. Commun. *1982*, 543
26. Wong, H. N. C., Luh, T.-Y., Mak, T. C. W.: Acta Crystallogr. *C40*, 1721 (1984)
27. Herbstein, F. H., Mak, T. C. W., Reisner, G. M., Wong, H. N. C.: J. Incl. Phenom. *1*, 301 (1984)
28. Chan, T.-L., Mak, T. C. W., Trotter, J.: J. Chem. Soc., Perkin Trans. 2 *1980*, 672
29. Weber, E., Csöregh, I., Stensland, B., Czugler, M.: J. Am. Chem. Soc. *106*, 3297 (1984)
30. Figeys, H. P., Dralants, A.: Tetrahedon Lett. *1971*, 3901
31. Barton, J. W., Whitaker, K. E.: J. Chem. Soc.(C) *1967*, 2097
32. Hellwinkel, D., Reiff, G., Nykodym, V.: Liebigs Ann. Chem. *1977*, 1013
33. Naito, K., Rickborn, B.: J. Org. Chem. *45*, 4061 (1980)
34. Man, Y. M., Wong K. C., Wong, H. N. C.: to be published
35. Hellwinkel, D., Reiff, G.: Angew. Chem. *82*, 516 (1970); Angew. Chem. Int. Ed. Engl. *9*, 527 (1970)
36. Hellwinkel, D., Haas, G.: Liebigs Ann. Chem. *1979*, 145
37. Thulin, B., Wennerström, O.: Tetrahedron Lett. *1977*, 929
38. Leach, D. N., Reiss, J. A.: J. Org. Chem. *43*, 2484 (1978)
39. Irngartinger, H., Reibel, W. R. K., Sheldrick, G. M.: Acta Crystallogr. *B37*, 1768 (1981)

40. Erdtman, H., Högberg, H.-E.: J. Chem. Soc., Chem. Commun. *1968*, 773
41. Högberg, H.-E.: Acta Chem. Scand. *26*, 309 (1972)
42. Erdtman, H., Högberg, H.-E.: Tetrahedron Lett. *1970*, 3389
43. Högberg, H.-E.: Acta Chem. Scand. *26*, 2752 (1972)
44. Kauffmann, T.: Angew. Chem. *91*, 1 (1979); Angew. Chem., Int. Ed. Engl. *18*, 1 (1979)
45. Greving, B., Woltermann, A., Kauffmann, T.: Angew. Chem. *86*, 475 (1974); Angew. Chem., Int. Ed. Engl. *13*, 467 (1974)
46. Kauffmann, T., Greving, B., Kriegesmann, R., Mitschker, A., Woltermann, A.: Chem. Ber. *111*, 1330 (1978)
47. Kauffmann, T., Greving, B., König, J. Mitschker, A., Woltermann, A.: Angew. Chem. *87*, 745 (1975); Angew. Chem., Int. Ed. Engl. *14*, 713 (1975)
48. Kauffmann, T., Muke, B., Otter, R., Tigler, D.: Angew. Chem. *87*, 746 (1975); Angew. Chem., Int. Ed. Engl. *14*, 714 (1975)
49. Kauffmann, T., Otter, R.: Chem. Ber. *115*, 1825 (1982)
50. Kauffmann, T. Otter, R.: Angew. Chem. *88*, 513 (1976); Angew. Chem., Int. Ed. Engl. *15*, 500 (1976).
51. Kauffmann, T. Otter, R.: Chem. Ber. *116*, 980 (1983)
52. Wong, H. N. C., Mak, T. C. W.: to be published
53. Chan, T.-L., Huang, N. Z., Sondheimer, F.: Tetrahedron *39*, 427 (1983)
54. Wong, H. N. C., Sondheimer, F.: J. Org. Chem. *45*, 2438 (1980)

Author Index Volumes 101–140

Contents of Vols. 50–100 see Vol. 100
Author and Subject Index Vols. 26–50 see Vol. 50

The volume numbers are printed in italics

www.ingramcontent.com/pod-product-compliance
Lightning Source LLC
LaVergne TN
LVHW060622300926
849162LV00008B/1685